Natural Gas

Resources Series

Natural Gas

Michael Bradshaw and
Tim Boersma

polity

First published in 2020 by Polity Press

Polity Press
65 Bridge Street
Cambridge CB2 1UR, UK

Polity Press
101 Station Landing
Suite 300
Medford, MA 02155, USA

ISBN-13: 978-0-7456-5997-8
ISBN-13: 978-0-7456-5998-5 (pb)

A catalogue record for this book is available from the British Library.

Library of Congress Cataloging-in-Publication Data

Names: Bradshaw, Michael J. (Michael John), 1935- author. | Boersma, Tim, author.
Title: Natural gas / Michael Bradshaw and Tim Boersma.
Description: Cambridge ; Medford, MA : Polity Press, 2020. | Series: Resources series | Includes bibliographical references and index. | Summary: "A compelling analysis of natural gas geopolitics"-- Provided by publisher.
Identifiers: LCCN 2019050902 (print) | LCCN 2019050903 (ebook) | ISBN 9780745659978 (hardback) | ISBN 9780745659985 (paperback) | ISBN 9781509542857 (epub)
Subjects: LCSH: Gas industry. | Gas industry--Political aspects. | Natural gas.
Classification: LCC HD9581.A2 B73 2020 (print) | LCC HD9581.A2 (ebook) | DDC 338.2/7285--dc23
LC record available at https://lccn.loc.gov/2019050902
LC ebook record available at https://lccn.loc.gov/2019050903
Typeset in 11 on 14 pt Sabon by
Servis Filmsetting Ltd, Stockport, Cheshire
Printed and bound in Great Britain by CPI Group (UK) Ltd, Croydon

For further information on Polity, visit our website: politybooks.com

Contents

Abbreviations

ACER	Agency for the Cooperation of Energy Regulators
APEC	Asia-Pacific Economic Cooperation
APR	Asia-Pacific region
bcf	billion cubic feet
bcm	billion cubic metres
bcma	billion cubic metres per annum
BECCS	biomass and CCUS
BGR	German Federal Institute for Geosciences and Natural Resources
Btu	British thermal unit
CAC	Central-Asia-Centre pipeline
CBM	coal bed methane
CCGT	Combined-Cycle Gas Turbine
CCS	carbon capture and storage
CC(U)S	carbon capture utilization and storage
CNG	compressed natural gas
CNPC	China National Petroleum Corporation
CSG	coal seam gas
EIA	(US) Energy Information Administration
ENR	Bureau of Energy Resources (US State Department)
EPA	(US) Environmental Protection Agency
ETS	Emissions Trading System
EU	European Union
FID	final investment decision
FLNG	Floating Liquefied Natural Gas

FSRU	Floating Storage and Regasification Unit
GGFR	Global Gas Flaring Reduction Partnership
GHG	greenhouse gas
GIIGNL	The International Group of Liquefied Natural Gas Importers
GTS	Gasunie Transport Services BV
IEA	International Energy Agency
IMO	International Maritime Organization
IOC	international oil company
IPCC	Inter-Governmental Panel on Climate Change
IGU	International Gas Union
LNG	liquefied natural gas
LPG	liquefied petroleum gas
mcf/d	million cubic feet per day
mcm/d	million cubic metres per day
MENA	Middle East and North Africa
MMBTu	million British thermal units
MTPA	million tonnes per annum
NBP	National Balancing Point
NGL	natural gas liquid
NGO	non-governmental organization
NIMBY	Not in My Back Yard
NOC	national oil company
OECD	Organisation for Economic Co-operation and Development
OIES	Oxford Institute for Energy Studies
ONGC	Oil and National Gas Corporation Limited (India)
PGNiG	Polskie Górnictwo Naftowe i Gazownictwo (Polish State Oil and Gas Company)
ppm	parts per million
SMR	steam methane reforming
TANAP	Trans-Anatolian Natural Gas Pipeline
TAP	Trans-Adriatic Pipeline

TAPI	Turkmenistan–Afghanistan–Pakistan–India Pipeline
tcf	trillion cubic feet
tcm	trillion cubic metres
TNK	Tyumen Oil Company (Russia)
TSO	transmission system operator
TTF	Title Transfer Facility
UKERC	United Kingdom Energy Research Centre

Natural gas (NG) and LNG conversion table

From	To					
	billion cubic metres NG	billion cubic feet NG	million tonnes oil equivalent	million tonnes LNG	trillion British thermal units	million barrels oil equivalent
	Multiply by					
1 billion cubic metres NG	1.000	35.315	0.860	0.735	34.121	5.883
1 billion cubic feet NG	0.028	1.000	0.024	0.021	0.966	0.167
1 million tonnes oil equivalent	1.163	41.071	1.000	0.855	39.683	6.842
1 million tonnes LNG	1.360	48.028	1.169	1.000	46.405	8.001
1 trillion British thermal units	0.029	1.035	0.025	0.022	1.000	0.172
1 million barrels oil equivalent	0.170	6.003	0.146	0.125	5.800	1.000

Source: BP, BP Statistical Review of World Energy, June 2019. London: BP, 2019.

Introduction

The global natural gas industry has experienced an unprecedented period of growth and change, best captured by the International Energy Agency (IEA)'s 2011 notion of the 'Golden Age for Gas'. When this book was first planned, it was conceived as an antidote to the tendency to lump the gas industry together with oil when considering geopolitics and energy security. We maintain that the materiality of natural gas as an energy source, and thus its geopolitics, is quite different from that of oil. Even today, the majority of natural gas is consumed within the country where it is produced. Until relatively recently, the vast majority of traded natural gas was moved by pipelines – it is only in the last decade or so that the growth of liquefied natural gas (LNG) production and trade has taken flight, to the extent that it will likely rival international pipeline trade sometime in the foreseeable future. However, most natural gas is still traded on a regional basis with regional prices and differences in price formation. In Asia, natural gas prices largely remain indexed to the oil price, although that may change due to market pressures, and institutional reform in key gas-consuming countries. In North America, and to a large extent in the EU, gas prices are formed based on gas-on-gas competition and the interplay between supply and demand. But, as this book explains, the status quo is continuously subject to change. Thanks to the expansion of the LNG trade, regional markets are increasingly linked and developments in one region impact on another. Thus, the price of natural gas in Europe is strongly influenced by the strength of demand in Asia (just as the oil price has traditionally played an important role). Investments in natural gas infrastructure are cyclical in nature. Periods of tight supply and high prices stimulate a new round of investment in production, which, depending on how demand responds to lower prices, may result in a period of over-supply and lower prices when new volumes come to market. As demand absorbs that supply and markets tighten, prices increase, stimulating a new round of investments, and so on.

Whether demand for natural gas will continue to grow is the subject of intense, and at times emotional, debate. As we explain, the natural gas industry has made much of the fact that it is the cleanest of fossil fuels, releasing half the amount of carbon dioxide (CO_2) emissions when burnt compared to coal, and having the ability to improve significantly the local air quality given the modest emissions of local air pollutants such as nitrous oxides ($NO_{x)}$ and sulphurous oxides ($SO_{x)}$. But the main component of natural gas – methane (CH_4) – is itself a potent greenhouse gas (GHG), and leakage of methane has the potential to cloud its aforementioned credentials.

At the same time, if we are to avoid the most catastrophic consequences of climate change, science dictates a transition to net zero emissions energy systems no later than mid-century. In this transition, natural gas has a role to play, but that role is different in different parts of the world and may be increasingly limited. Unless industry can improve the GHG footprint of natural gas (e.g. by curtailing CH_4 emissions, minimizing flaring, and investing in hydrogen, biomethane, and CCUS technologies), sooner or later the fuel will become part of the problem, rather than part of the solution. To be sure, a section of the environmental community has already reached that conclusion. In this context, our understanding of natural gas demand is blurred. We observe that most forecasts and scenarios expect natural gas to be the only fossil fuel experiencing demand growth well into the 2040s (and sometimes beyond). However, other scenarios and forecasts portray a different future for all fossil fuels, including natural gas. The IEA's Sustainable Development Scenario, for example, suggests that demand for natural gas grows until the 2030s, but then levels off and declines substantially. As described, there appears to be significant room to improve the GHG footprint of natural gas, and if industry were to succeed in doing so, one would expect numbers in various studies and outlooks to start shifting again (assuming net zero emissions policies, rather than fuel or technology prescriptions). In addition, natural gas seems poised to play a prominent role in the fuel mix of many emerging economies, where air quality concerns trump many other policy objectives (and, in many instances, this shift results in benefits in terms of GHG emissions reductions as well). Consequently, natural gas is likely to play a substantial role in energy transition pathways in many parts of the world – yet the further out we look, the larger the uncertainties grow.

As a result of the issues described above, this book has been written in the context of natural gas abundance, swiftly growing demand and growing global interconnectedness, on the one hand, and mounting uncertainties about the role of natural gas in future low-GHG energy systems, on the other. The structure of the book is built around the evolving impact of three recent, and ongoing, revolutions. The first is the shale gas revolution that has heralded an age of resource abundance in North America, resulting in a surge in gas consumption in the United States and the growth of both pipeline and LNG exports, with ripple effects around the globe. The second is the coming of age of the LNG industry that is mobilizing new gas reserves

– both conventional and unconventional – to satisfy growing demand in emerging economies that could until recently not tap into this market, providing a source of supply diversification in mature markets, and allowing emerging economies to monetize resources that were long considered difficult to bring to market. The third revolution – namely, improvement of the environmental footprint of the resource – is essential if natural gas is to play a significant long-term role in various energy transition pathways. This book is a modest effort to make accessible a set of complex and fast-changing issues. Inevitably, it is far from comprehensive and some readers will feel that key regions and issues have been omitted or only touched on lightly. We have chosen to focus mostly on those regions that are critical in shaping supply of and demand for natural gas as we currently observe it – namely, North America, Eurasia, the Middle East and East Asia. That is not to say that developments elsewhere are not important, but we believe that this trade is at the centre of the processes that are shaping supply and demand for natural gas.

To adapt a well-known African proverb, it has taken a village to complete this manuscript. For Michael Bradshaw, funding from the UK Energy Research Centre (UKERC) has been essential to supporting research on the global gas industry, as has funding from the EU's Horizon 2020 programme for a project on the shale gas industry. He has also benefitted greatly from involvement in the Gas Programme at the Oxford Institute for Energy Studies (hereafter OIES) – surely, the world's greatest concentration of 'gas geeks'. Over the time it has taken to complete this book, much has changed, and he would like to dedicate this book to the memory of his father, Joseph Bradshaw, and his mentor Peter Daniels. He would also like to thank his family for their love and support, with apologies for the weekends lost working on this and many other projects.

The work of Tim Boersma was made possible by generous support from the Center on Global Energy Policy. He would like to thank his colleagues at Columbia University, and those in the natural gas research programme specifically, for numerous stimulating conversations and joint research projects. As always, he thanks his 'partner in crime' Susana, and his family and friends for their continued love and support. This manuscript would not have come to fruition without them.

We thank Louise Knight at Polity Press who first commissioned the book and who has shown great patience and provided tremendous support throughout the process. Most recently, Inès Boxman has provided the support and cajoling necessary to get the book across the finish line. We also thank our reviewers for finding the time and lending their expertise to evaluate this work. As ever, any omissions and errors remain our responsibility.

Michael Bradshaw
Coventry

Tim Boersma
New York

CHAPTER ONE

Natural Gas Fundamentals

For most private and state-owned energy companies, natural gas is considered an affordable, accessible and acceptable route to a future low-carbon energy system. For most environmental groups, it is just another fossil fuel and, as such, is part of the problem, rather than the solution. A recent NGO (non-governmental organization) report maintains that gas is not clean, cheap or necessary.[1] A parallel narrative argues that the billions of dollars being invested in new natural gas production and infrastructure, that will take decades to pay off, could end up as stranded assets in a future world of significantly constrained GHG emissions. Between these opposing views lies the position that natural gas may in many, but not all, parts of the world provide a bridge to a future low-carbon energy system and, depending on technology and policy, may play a role thereafter.[2] This is why mainstream energy forecasts see natural gas as the only fossil fuel that will enjoy sustained growth in demand through the first half of this century.

Our belief is that the scale and pace of global gas demand growth will be determined by the complex interplay of economics, geography and (geo)politics within the wider context of energy and environmental (including climate change) policy. As explained in the introduction, we posit that the natural gas industry has been undergoing two revolutions in recent years, which have contributed to the swift rise of natural gas as a fuel and feedstock in most parts of the world. However, a third revolution is required for natural gas to cement its long-term role, as governments and business seek to curtail GHG emissions and limit environmental pollution in the decades ahead. This determination results in what we call a 'geopolitical economy approach' to analysing the global gas industry, the details of which are discussed at the end of this chapter. But before we can discuss the approach and content of this book in any detail, we start with the fundamentals of natural gas.

The what, how and where of natural gas

It is commonplace for geopolitical analysis to lump oil and gas together as if they were the same thing, but it is important to understand the material specificity of natural gas as this explains in large part why the geopolitics of natural gas are different and distinct from those of oil.[3] This section asks four questions about the materiality of natural gas that have significance for our analysis: what is natural gas, how does it occur, how is it measured and where is it found? In a recent study, Peter Evans and Michael Farina identified three eras of gas: the first was the era of 'manufactured gas' (or 'town gas') from 1810 to around 1920, when gas was produced primarily from coal; the second was the era of 'conventional gas' from 1920 to around 2000; and the third era of 'unconventional gas' is still unfolding.[4] However, we would emphasize that the age of conventional gas is far from over and what we are witnessing is the global expansion of the natural gas industry based on proven reserves and new discoveries of conventional gas and the promise of more geographically dispersed unconventional gas. In addition, there is a widely held, though contested, belief that natural gas will continue to play a role in a low carbon economy as well. This book focuses on developments since the 1920s and does not consider the era of manufactured gas.

What Is Natural Gas?
Natural gas terminology can be confusing, because definitions are often absent. The main component of natural gas is methane (CH_4), which tells us the chemical composition: one carbon element (C) and four hydrogen elements (H). If the percentage of methane in natural gas is very high (>95 per cent) it is called 'dry gas'. It will require little effort to process the natural gas once brought to the surface (more on this later) and is relatively easy to use. More often, the share of methane is more modest, and methane is found together with heavier molecules – in the industry lexicon, this is often lumped together as 'natural gas liquids' (NGLs). This is also referred to as 'wet gas'. These NGLs include ethane (C_2H_6), propane (C_3H_8) and butane (C_4H_{10}). When pressurized in containers, the latter two liquids are often referred to as liquefied petroleum gas (LPG). Whether a natural gas field produces 'wet' or 'dry' gas is an important commercial consideration as the NGLs are valuable in their own right. In fact, depending on market conditions, NGLs are regularly the main commercial incentive for producers to invest, with natural gas becoming an associated product that can be sold at a very competitive price. This is logical when considering that NGLs have a higher energy content and therefore typically have a higher value than methane.

Natural gas often also contains non-hydrocarbon elements. The most common ones are carbon dioxide, hydrogen sulphide, hydrogen and helium. Other impurities that may be present are water, sulphur species, mercury, naturally occurring radioactive materials and oxygen.[5] These elements are removed during processing. If they are encountered in too large quantities, it can be too costly to produce the natural gas (in which case the resource is left in the ground). Natural gas with a high hydrogen sulphide content, for example, is called 'sour gas', and is sometimes left where it is.[6]

The composition of natural gas determines the amount of energy it contains, which is released when the gas is completely burned. Most natural gas that is found around the world has a high heating value, or high calorific value. Natural gas meeting a certain bandwidth of specifications can therefore be used in appliances interchangeably. Variations in quality of up to 5 per cent typically do not affect usage of the natural gas in appliances. The Wobbe index is an indicator of the inter-changeability of fuel gases such as natural gas, LPG and town gas. In rare instances, natural gas has a low calorific value. The giant Slochteren field in the Netherlands, discovered in 1959 and producing to date, is the best example. Because the natural gas from this field is of a different quality, burners in house-hold and many industrial appliances have been aligned with this natural gas (known as L-Gas because of its lower calo-rific value). Other natural gas found in the Netherlands – for instance from smaller fields in the North Sea – and imported natural gas – e.g., from Norway, Russia, or in the form of LNG – tends to be of high calorific value (known as H-Gas). In order to be able to use this natural gas, the transmission system operator GTS (Gasunie Transport Services BV) operates a nitrogen plant. By mingling high-calorific gas with nitrogen, it can be inserted into the Dutch distribution grid, and used in household appliances. Major industrial consumers in the Netherlands are generally not connected to the low-calorific gas grid. As discussed in more detail in chapter 2, the phasing out of natural gas production in the Netherlands has major consequences for Dutch consumers, and those in Northern Germany, Belgium and Northern France (who also use low-calorific Slochteren gas). Their appliances need to be adjusted at the household level, in order to prepare for H-Gas, once production in the Netherlands ceases entirely, sometime in the early 2020s.

Once natural gas has been brought to the surface, it needs to be treated in order to have roughly similar specifications flowing through pipelines. For dry gas, impurities must be removed to get the natural gas at a quality level that allows transportation to markets and/or end users. For wet gas, NGLs are generally removed in order to sell those in their own markets. Treatment of natural gas takes place in a processing plant, sometimes described as a giant refrigerator.[7] When a stream of gas is cooled, NGLs condense into a liquid, allowing for separation and further treatment. The process to separate NGLs into ethane, propane and so on is called fractionation, and occurs in a separate processing plant. Once separation has taken place, natural gas can be injected into high-pressure transmission lines for transport to markets. Once closer to end users – industries and households – natural gas is transported in low-pressure distribution grids at the local level. Some large consumers, e.g. electricity or industrial plants, take natural gas directly from the transmission pipeline. Of course, in many parts of the world, infrastructure to transport natural gas to market has not been developed sufficiently, if at all. Demand can be scattered, making pipelines uneconomic to build, and in many developing countries the risks of making capital-intensive investments against the backdrop of uncertainties related to economic development or governance are just too large. Absent transport by pipelines, natural gas can be cooled until it becomes a liquid – so-called liquefied natural gas (LNG) – and transported by ship (and even by container and truck) to reach end users. As described in detail in chapter 4, the market for LNG has been around since the 1960s, but is currently undergoing fundamental change and is growing rapidly.

When combusted (burned), natural gas releases carbon dioxide (CO_2), the largest contributor to anthropogenic climate change. If released into the atmosphere without combustion,

CH_4 has a different, but also very significant, impact on the earth's climate. In its Fifth Assessment Report, the Inter-Governmental Panel on Climate Change (IPCC) upgraded the global warming effects of methane to 34 times stronger than CO_2 over a 100-year period, up from 25 times in their previous report in 2008.[8] Over a 20-year period, the effect is now rated as 86 times stronger than CO_2. This is significant because it means that any leakage (known as fugitive emissions) of methane during the production, processing, transportation and consumption of natural gas makes a significant contribution to climate change on top of the CO_2 released as a result of combustion. As we discuss in more detail later, our collective understanding of fugitive emissions throughout the production cycle is relatively poor, though receiving more attention recently, but it is safe to say that there is significant room to curtail fugitive methane in industry practices. Furthermore, given its global warming potential, it is preferable to burn or flare off methane when produced as a by-product in the production and processing of oil as this results in CO_2 being emitted into the atmosphere, rather than methane. However, flaring is itself a significant contributor to climate change and is also a waste of a valuable hydrocarbon. The World Bank has a leadership role in gas-flaring reduction through the Global Gas Flaring Reduction Partnership (GGFR), an international programme to reduce the level of gas flaring to zero by 2030.[9] The World Bank assesses that flaring emits about 400 million tons of CO_2 into the atmosphere each year, about 1.2 per cent of total global emissions; in 2018, the World Bank estimated the total volume of flared gas to be 145 bcm. Since 2002, the practice had generally been in decline, yet the amount of natural gas that was flared off grew by 3 per cent in 2018, in no small part as a consequence of the rise of oil production in the United States.[10]

Most of the natural gas that is commercially produced today is thermogenic in origin: it is the product of the application of heat and pressure over geologic time to organic matter that originally came from plants and animals and has since become buried thousands of metres underground. By contrast, biogenic natural gas is formed by microbial action on organic matter in an anaerobic (without oxygen) environment, and can be produced in shallow formations, swamps and landfill. The same process is used in an anaerobic digester that uses food or agricultural waste to produce biogas. Many of the images in the movie *Gasland* showing kitchen taps on fire turned out to be of biogenic origin or a result of earlier coal mining, and not from shale-gas activity. Thermogenic methane can be found naturally occurring in aquifers – so-called aquifer gas – but poor well completion can also pollute water supplies, and is as much a problem for conventional oil and gas wells as for shale-gas production.

How Does Natural Gas Occur?
Interest in the different ways in which natural gas occurs in the sub-surface has been heightened by the development of so-called 'unconventional gas'. Figure 1.1 classifies different types of natural gas on the basis of their geological mode of occurrence, the rate at which gas flows within the source rock, and the level of technology and the amount of work required to extract that gas (net energy return on investment).[11] Conventional gas is found in two forms; the first, associated gas, is found in conjunction with oil or NGLs, either mixed with the oil or in a separate gas cap. Historically, the oil industry considered this natural gas a complication as they were most interested in producing crude oil, and if the associated gas was not marketable, or could not be used to enhance oil production, it was flared off. As noted above, the practice con-

tinues today. The tight oil fields in the US in North Dakota and Montana made the press because so much gas was being flared off that satellite images suggested that a new metropolis had suddenly emerged in the Mid-West.[12] Absent local markets to speak of, infrastructure to bring natural gas to market had not been developed and consequently most natural gas was flared off; since then, regulatory authorities and industry have made a concerted effort to expand processing and transport capacity, in part fuelled by societal concerns that a valuable resource was being wasted. More recently, attention has turned to the Permian Basin in western Texas, where the same problem has developed as oil production has surged. The amount of natural gas flared in that part of the country has risen dramatically, surpassing 340 mcf/d (million cubic feet per day; this is equivalent to 14.2 mcm/d – million cubic metres per day) in September 2018,[13] and possibly up to 1 bcf/d (billion cubic feet per day; 28.3 mcm/d) in 2019, as more takeaway capacity is being constructed. Pioneer CEO Scott Sheffield called the enormous amounts of natural gas being flared in West Texas – enough to power every home in Texas – a 'black eye' for the industry at an energy conference in New York in April 2019.[14] However, others see the availability of large volumes of cheap gas in close proximity to the Gulf of Mexico as a major opportunity, provided that the necessary pipeline and LNG export infrastructure can be built (this is discussed further in chapter 4).

The second form of conventional gas is 'non-associated' and, as the name implies, this is where natural gas is present without oil. In both instances, the conventional gas is found in well-defined accumulations known as reservoirs. The permeability of the reservoir – the ability of the rock to transmit fluids – varies by rock type and is measured in darcies or millidarcies (after the French engineer Henry Darcy). Conventional

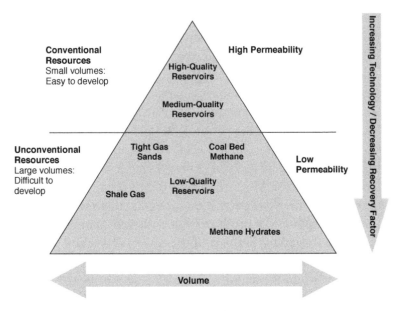

Figure 1.1   Natural gas resource triangle

reservoirs are sufficiently permeable to enable the gas to flow freely and it can be extracted under its own pressure via a vertical well. The gas is held in place by an impermeable cap rock. It is important to note that the practice of hydraulic fracturing has long been used in conventional oil and gas reservoirs to enhance production. The term 'unconventional gas' commonly refers to the way in which the gas molecules are held within the fabric of the rock, rather than the techniques used to extract it.

Figure 1.1 is organized in terms of declining permeability and increasing occurrence. Thus, the bottom half of the triangle is occupied by increasing low-permeability gas resources that are widely available across the globe. The precise definition of the different types of unconventional gas is shown in table 1.1. The key distinction is that, because of their very low permeability, it is necessary to break up or fracture the rock to release the gas molecules that then flow through the fractures that are held open by grains of sand or similar material known as proppants (more on this in chapter 3). Tight gas is found in sandstones and limestone, while shale gas – as the name suggests – is found in shale rock. Coal bed methane (CBM), also known in Australia as coal seam gas (CSG), describes gas which is trapped in coal seams. Historically, methane has presented a major safety hazard to the coal-mining industry. As an early detection mechanism, coal miners used to bring caged canaries into the mines with them, which, if there was a leak, would die before the concentration of gas reached levels hazardous to humans. However, natural gas can be produced from coal seams that are either too deep and/or of insufficient quality to warrant mining. The production of CBM involves first pumping out the ground water in the coal seams to release pressure and promote gas flow, and, if necessary, hydraulic fracturing is then used to create a pathway for the gas to flow. This process should not be confused with the far less common practice of underground coal gasification that involves the combustion of the coal to produce 'syngas'. The final category of natural gas in figure 1.1 is gas hydrates, which are an ice-like crystalline form of water and gases – including methane – that are found in some marine sediments and within and beneath permafrost (permanently frozen ground). These are really a category of their own and relatively little is known about the volume of methane stored in gas hydrates, though it is very large.[15] As commercial exploitation does not seem likely any time soon, they are not considered in this analysis.

Before we conclude this discussion, there are two further factors to consider: the first is the amount of energy that is consumed in the production of these various forms of natural gas; and the second is the cost of development. It was noted earlier that unconventional gas resources require additional

---

Table 1.1 Natural gas: definitions

Natural gas: gas occurring naturally underground or flowing out at the surface. Gases can have variable chemical compositions but in this context are understood to be combustible natural gases.

Wet gas: contains methane (CH_4) as well as longer-chain hydrocarbon constituents often known as natural gas liquids or NGL: ethane (C_2H_6), propane (C_3H_8) and butane (C_4H_{10}). Gas fields with high levels of NGL are often referred to as Gas Condensate Fields.

Dry gas: only contains gaseous components and mainly consists of methane: over 95 per cent.

Sour gas: contains varying amounts of hydrogen sulphide ($H2_S$) in the ppm range.

Conventional natural gas: natural gas or associated gas in structural or stratigraphic traps.

Non-conventional gas: due to the nature and properties of the reservoir, the gas does not usually flow in adequate quantities into the production well without undertaking additional technical measures – includes shale gas, tight gas, coal bed methane, aquifer gas, and gas from gas hydrates.
  Shale gas: natural gas from fine-grained rocks (shales).
  Tight gas: natural gas from tight sandstones and limestones.
  Coal bed methane or CBM: gas contained in coal seams.
  Aquifer gas: natural gas dissolved in ground water.
  Gas hydrates: solid (ice-like) molecular compound, consisting of gas and water, which is stable under high pressure and low temperatures.

Source: based on BRD (2014), *Energy Study: Reserves, Resources and Availability of Energy Resources.* Berlin: Federal Institute for Geosciences and Natural Resources (BGR).

work to extract the gas from the source rock. In simple terms, more energy must be put in to recover an equivalent amount of natural gas (energy), compared to conventional resources. This means that the net return on energy invested – the difference between the energy invested in extraction and the energy available in the produced natural gas – is lower for unconventional gas. Furthermore, generally speaking, the rate of production from an unconventional gas well declines more rapidly than is the case with conventional wells. The net result is that the recovery rate of unconventional gas – the percentage of the gas in place that is produced over the lifetime of the well – is lower. However, what is critically important is that, although it has proven tempting for analysts to draw definitive conclusions about shale, this is a relatively young industry which is maturing in what is the most competitive energy landscape in the world – namely, the United States. It is therefore inevitable that there is still much to learn, for both analysts and industry, as practices mature and improve. A case in point is the improvement in recovery rates from wells in recent years.[16] It is also important to acknowledge the significant rise in associated natural gas production, and how this has contributed to the consistent growth of US natural gas output over the last decade, even though most drilling rigs in the country were used to target oil (and NGLs).

The discussion above suggests that there is a clear relationship between the cost of production of the different types of natural gas and the amount of gas that can be brought to market at a given price. Simply put, the higher the price, then the more gas that can be produced. The flip side is that if the gas price falls for an extended period of time, certain types of gas production can become unprofitable.[17] Of course, technology, costs and prices are not static, and do not evolve in a vacuum either. Thus, analysts must be mindful of this when forecasting the future with too much certainty. The relationship between technology and price is commonly illustrated in supply curves that illustrate how much natural gas would be produced at a given price level. Though regularly used interchangeably, costs and prices are very different things. As a 2018 Columbia University study illustrated, LNG producers such as Qatar can produce natural gas at a low, close to zero, cost (because the North Field from which it is produced is very rich in NGLs).[18] However, despite a *Financial Times* commentary suggesting otherwise, this does not inform us about the price that Qatargas charges its clientele, which will depend on market prices of competing natural gas, other feedstocks and demand.[19]

How Is Natural Gas Measured?
The discussion above has already introduced the difficult issue of how natural gas is measured, and reserves and production reported. In simple terms, natural gas is measured by volume and by energy content. The problem of measurement is further complicated by the existence of both imperial and SI units (metric units). The imperial measure of volume is cubic feet and the SI equivalent is cubic metres. Large volumes, such as annual production and consumption figures, are measured in trillion cubic feet (tcf) or billion cubic metres (bcm), while reserves are reported in trillion cubic feet and trillion cubic metres (tcm). The US government's Energy Information Administration (EIA) and the US energy industry continue to use imperial units, while the rest of the world favours SI units or a confused mixture of both. To complicate matters even further, there is even variation in the amount of gas (energy) in a cubic metre. In Russia, a cubic metre is measured at a temperature of 20 °C and a pressure of 760 mm of mercury (Hg), while the European standard is at 15 °C and 760 mmHg. When it comes to measuring the energy content of natural gas, the imperial measure is the British thermal unit (Btu), which is the amount of heat that is required to cool or heat 1 pound of water by 1 degree Fahrenheit. The SI equivalent is the Joule (J), which is the heat required to raise the temperature of 1 gram of water by 1 degree Celsius at standard pressure. A Joule is a small unit of energy, and in Australia public discussion of natural gas demand is often in terms of petajoules (a quadrillion Joules). One petajoule is sufficient to supply a fair-sized town for a year. To compare the two, 1 Btu is equivalent to 1,055 Joules; 1 cubic foot of natural gas has approximately the heating capacity of 1,000 Btus, and a therm is 100,000 Btus. To confuse matters further, LNG production is also measured by weight, usually in terms of millions of tonnes per annum (MTPA).

When it comes to the price of natural gas, this is also quoted in a number of different ways, the most common being US dollars per million Btu ($/MMBtu); prices are also quoted in US dollars per therm ($ per therm), dollars per 1,000 cubic metres ($/mcm), and dollars per 1,000 cubic feet ($/mcf). In Continental Europe, euros per Megawatt hour (euro/MWh) is the most common unit, and in the UK, pence per therm (p/therm). Conversion between pricing systems based on different currencies is subject to movements in exchange rates that impact on the availability and affordability of natural gas. For example, the devaluation of the pound was one of the most immediate consequences of the UK's Brexit vote in 2016, and has increased the cost of gas imported in US dollars and euros. Ideally, one would be able to standardize the measurements used, but this is complex and time-consuming. In this analysis, we will use the most commonly used measurements, which is a mix of imperial measures for value, and SI measures for volume. A conversion table is provided at the beginning of the book, and there are online calculators that make individual calculations relatively straightforward.[20]

Where Are Natural Gas Resources Located?

The final question to be answered in this section is the 'where' of natural gas: which countries have the greatest reserves and resources? Unfortunately, this is yet another complex question as much depends on what you are measuring. The key distinction is between resources and reserves, the former being an estimate of the amount of natural gas in the ground, and the latter being the amount of natural gas that can be technically and commercially produced today. Table 1.2 provides information on the distribution of natural gas resources and reserves by major world region, and also includes the major reserve-holding countries. A further complication here is that different organizations divide the world up in different ways, which makes comparison between data from the IEA *World Energy Outlook*, the *BP Statistical Review of World Energy* and the EIA's *International Energy Outlook* difficult, to say the least.[21] Table 1.2 is based on data from the German BGR and provides definitions for all of the categories used. A large part of the confusion over the difference between resources and reserves comes from the estimates for the global distribution of unconventional gas resources, most of which are not based on extensive exploration programmes.

The starting point in the assessment of shale gas reserves, for example, is a geological appraisal of the gas in place. The British Geological Survey has produced an estimate of the amounts of oil and gas they believe to be physically contained in the source rock – in this case, the Bowland–Hodder shale formation.[22] Because of the high level of uncertainty, they produced a range of estimates, and their central estimate is that there is 37.6 tcm of gas in place across the Bowland–Hodder shale basin. However, this is not an estimate of proven reserves of natural gas, which according to BP are: 'those quantities that geological and engineering information indicates with reasonable certainty can be recovered in the future from known reservoirs under existing economic and operating conditions'. To arrive at this number requires increased exploratory drilling and geological appraisal. The US EIA has produced a set of global shale gas estimates for technologically recoverable resources (TTR), which represents the volume of natural gas that could be produced with current technology, regardless of oil and natural gas prices and production costs.[23]

As Table 1.2 demonstrates, the largest numbers are for resources and the smaller numbers are reserves. As technology and price levels change, so do resource and reserve estimates. The so-called 'shale gas revolution' is so significant because,

Table 1.2 Natural gas: resources, reserves and production, 2016 (bcm)

	Production	Cum. production	Reserves	Resources	EUR	Remaining potential	Remaining potential %
Europe	253.2	12,815	3,229	19,316	35,361	22,545	2.7
CIS	838.1	31,300	63,262	179,114	273,676	242,375	28.9
(Russia)	*640.7*	*22,966*	*47,777*	*152,050*	*222,793*	*199,827*	*23.7*
Africa	206.7	4,572	14,377	79,739	98,688	94,116	11.2
Middle East	629.4	9,410	79,370	56,074	144,854	135,444	16.1
(Iran)	*202.4*	*2,766*	*33,721*	*10,000*	*46,487*	*43,721*	*5.2*
(Qatar)	*165.4*	*1,772*	*24,073*	*2,000*	*27,844*	*26,073*	*3.1*
(Saudi Arabia)	*109.4*	*2,005*	*8,427*	*24,664*	*35,096*	*33,091*	*3.9*
Austral-Asia	564.8	10,689	17,569	131,682	159,939	149,250	17.8
(China)	*141.9*	*1,777*	*5,191*	*64,900*	*71,868*	*70,091*	*8.3*
North America	960.1	43,820	11,155	112,817	167,793	123,973	14.8
(United States)	*755.8*	*35,806*	*8,714*	*53,246*	*97,766*	*61,960*	*7.4*
Latin America	167.5	4,280	7,643	64,416	76,339	72,059	8.6
WORLD	**3,619**	**116,886**	**196,605**	**643,157**	**956,648**	**839,762**	**100.0**

Definitions

Italics indicate a major reserve-holder in the relevant region.

CIS: Commonwealth of Independent States.

Cumulative production: total production since the start of production operations.

Reserves: the proven volume of natural gas that is economically exploitable at today's prices and using today's technology.

Resources: proven amounts of natural gas that cannot currently be exploited, for technical and/or economic reasons, as well as unproven, but geologically possible, energy resources that may be exploitable in the future.

EUR: the Estimated Ultimate Recovery is the total amount of natural gas that can be extracted from a deposit.

Remaining potential: reserves plus resources.

Source: adapted from BGR (2016), *Data and Developments Concerning German and Global Energy Supplies.* Berlin: BGR.

by harnessing a combination of established technologies, it has greatly increased both the resource base and potentially the reserve base on a global scale. The data in table 1.2 include a modest amount of unconventional gas, most of which is in North America. The bottom line is that there is still a lot of natural gas available, with the prospect of even more being confirmed in the future. However, the geography of current conventional reserves is different from that of potential future unconventional reserves. Over half of conventional natural gas reserves are concentrated in three countries: Russia, Iran and Qatar. By contrast, the EIA's 2014/15 estimates suggest that technically recoverable shale gas resources are more evenly spread, with the top three – China, Argentina and Algeria – accounting for just over a third of total resources. These estimates and the shale gas revolution are examined in detail in chapter 3; for the moment, we can conclude that the natural gas resource base – conventional and unconventional – is substantial and widely geographically distributed; however, proven reserves of conventional gas are currently more geographically concentrated.

International trade in natural gas

Compared to oil, natural gas is a higher-volume and lower-value resource; this makes it relatively expensive to transport per unit of energy. The most economical way to transport natural gas is using pipelines. The gas is compressed and pushed through the pipe: the larger the diameter of the pipe and the higher the level of compression, the more gas is delivered to market. There are pipelines that gather natural gas from wells and take it to processing facilities, and feeder pipelines then connect it to longer-distance transmission pipelines that evenly connect to a distribution pipeline network that delivers gas to consumers. The different types of pipe are of different diameters and pressures, making it necessary to balance the system. Some long-distance pipelines travel thousands of kilometres delivering gas – for example, from fields above the Arctic Circle in Siberia to customers across Europe. The cost of large-scale pipelines that move gas over thousands of kilometres can range from $1.50 to $3.75 per MMBtu.[24] While pipelines are an efficient way of moving large volumes of gas, they are expensive to build and, once in place, they are inflexible and tie producers and consumers into a long-term interdependent relationship that can be vulnerable to (geo)political manipulation (this is the subject of chapter 2). A final point to consider is that pipelines and their associated infrastructure consume energy and generate carbon emissions; they are also leaky – to varying degrees, depending on age and upkeep – thus the transport of pipeline gas is a source of GHG that needs to be considered in the life-cycle emissions associated with natural gas consumption.[25]

The alternative way to transport natural gas is to cool it and turn it into LNG, which takes up 600 times less space than its gaseous form. This requires a considerable amount of energy and a capital-intensive supply chain comprising a liquefaction plant, specialized ships and a regasification terminal to unload the LNG, store it and then return it to the gaseous state before connecting it to a pipeline system or combusting it in applications. All of these stages, but in particular liquefying the natural gas since this is an energy-intensive process, add additional cost (as a rule of thumb, liquefaction costs can range from 0.9 to 1.3 $/mcf, transport from 0.40 to 1.0 $/mcf depending on distance, and regasification from 0.35 $/mcf). Still, LNG has historically been an attractive fuel for island states, and regions where the necessary pipeline infrastructure has not been built. It is now emerging as an option for countries that do not wish to invest in costly pipeline systems. For reasons that are explained in chapter 4, historically the majority of LNG trade was conducted on the basis of long-term contracts that tied suppliers and buyers to a schedule of agreed deliveries at predetermined points. However, as we discuss later, various significant changes are underway in the market for LNG, increasing its attractiveness as a flexible fuel, leading to a significant expansion of the market in recent years. Finally, natural gas can also be transported as compressed natural gas (CNG) and LNG by truck, train and container.[26]

The challenges of transporting natural gas also explain two further ways in which gas differs from oil: first, that the majority of natural gas production is actually consumed in the producing region; and second, that natural gas is organized around a series of regional markets, rather than a global market as is the case for oil, even though with the growth of trade by ship regional markets seem, albeit slowly and haphazardly, to be becoming more interconnected than they have historically been.[27]

Our demand for energy is rarely constant, thus creating a value to storing it, if possible. Natural gas is no different. In fact, given the significant proportion of natural gas used in heating, weather patterns can have a major impact on natural gas demand, particularly in the northern-hemisphere winter as the major demand centres are North America, Europe and Northeast Asia. Historically, one of the most important roles performed by natural gas storage was to provide a cushion for seasonal swings in natural gas demand, which can be witnessed in most advanced economies. Typically, demand for natural gas – for heating or incremental electricity generation – picks up in October and remains elevated throughout the winter months. Demand is typically muted in the shoulder months during the spring, and – depending on local cooling demand – it can have a second peak during the northern-hemisphere summer, particularly in Asia, as air conditioning drives greater electricity use. If we assume stable production of natural gas from producing wells, then we would expect a downward effect on natural gas prices over the summer, and, fuelled by incremental demand, an upward effect in winter. This spread between summer and winter prices has historically been a key incentive for operators of natural gas storage facilities. Seasonal storage facilities are typically filled during the spring and summer season, when demand and prices are the lowest, and withdrawals begin around October, when demand and prices pick up. In broad strokes, this is the commercial model behind natural gas storage.

As well as covering seasonal swings in demand, natural gas storage can serve several additional purposes. For example, natural gas storage can serve more sudden peak demand (think, for instance, of a short cold snap in the winter). However, not all storage facilities are equally equipped to serve this purpose. Natural gas can be stored underground in depleted oil and gas reservoirs, aquifers or salt caverns. Depleted reservoirs are the least expensive and most common way to store natural gas. Because there are limitations regarding the amount of natural gas that can be extracted in a short amount of time, this type of storage is mostly suited to meeting seasonal demand swings. Storage of natural gas in aquifers is less common and more costly (due to the high cushion gas requirement, which is the minimum volume of gas required in an underground storage reservoir to provide the necessary pressure to deliver working gas volumes to customers), and it faces similar challenges in terms of send-out capacity. Salt caverns require little cushion gas, which allows very high injection and withdrawal rates within a short period of time. This type of gas storage can meet short-term (or even intra-day) swings in demand, which other types of storage facilities are unable to do.

Natural gas storage can also help to support transport flexibility (at times when demand is difficult to predict), or support trading flexibility (consider, for instance, that traders operate in different time zones all around the world, and trading patterns may therefore be difficult to predict). Then, finally, natural gas storage can serve strategic purposes, most prominently in countries that depend on imported natural gas, and are concerned about potential geopolitical or technical disruptions of their gas supply. In those cases, the underlying question is, of course: how certain do companies and governments want to be that they have an uninterrupted supply of natural gas? This question is, for instance, prevalent in the United Kingdom, where Centrica in 2017 decided to close its Rough storage facility. With a total capacity of 3 bcm, closing this facility has meant reducing the available storage capacity by almost 75 percent (total capacity before the announced closure was 4.3 bcm). Centrica concluded that it could no longer operate this facility on a commercial basis, because the cost of maintaining it was too high and because the spread between summer and winter prices has evaporated. Thus, the UK now has to rely on other forms of flexibility available in the market – for example, remaining storage capacity, interconnection to the continental European gas market, and the availability of lots of natural gas in the form of LNG in the global system. There is no business case for building new gas storage in the UK today, and for British policymakers and regulators the key question seems to revolve around the systemic value of natural gas storage, and whether its energy-security benefits justify the socialization of its costs at a time when the economics of commercial storage is under pressure from other forms of gas supply flexibility. In the context of several other sources of flexibility being taken out of the Northwest European market – specifically the Netherlands, as we discuss later – we can expect to hear more about natural gas storage in the European Union (EU) in the years ahead.

A Global Industry with Regional Markets

According to the *BP Statistical Review of World Energy*,[28] in 2018 total global natural gas production was 3,867.9 bcm, while the total volume of trade movements by pipeline was 805.4 bcm and the total for trade movements as LNG was 393.4 bcm. This means that internationally traded gas accounted for 32% of total gas production in 2018. Of that traded gas, 65.1% was transported by pipeline and 34.9% by ship as LNG. In 2008, the total volume of global gas production was 3045.4 bcm, of which 26.7% was traded, 72.2% of that was by pipeline and 27.8% by ship as LNG. Thus, there is a trend over time of an increased share of global production being internationally traded, with LNG gaining an increasing share of that trade. Whether these trends are set to continue is a question to be addressed in the conclusion.

The geographies of natural gas production and consumption provide the logic for the pattern of international trade. In 2018, natural gas production was concentrated in North America, the former Soviet Union and the Middle East. The leading producer was the United States (831.8 bcm), followed by Russia (669.5 bcm), Iran (239.5 bcm), Canada (184.7) and Qatar (175.4 bcm). These five producers accounted for 54.3% of global production. By comparison, the largest consumers were the United States (817.1 bcm), the EU (458.5 bcm), Russia (454.5 bcm), China (283.0) and Japan (115.7). These top five consumers accounted for 55.3% of global consumption. North America (including Mexico) is essentially a self-contained market that is now largely self-sufficient (with large volumes of US gas flowing into Mexico). The growth of gas production in the US is a major problem for producers in Western Canada, who find that their export market has become a source of competition, and they are struggling to develop access to markets in Asia and Europe. The EU is increasingly dependent on imports, the majority of which come by pipeline from Norway, Russia and North Africa. Starting in late 2018 and into the first half of 2019, LNG imports into the EU were up, though that may change going forward depending on a variety of factors, such as availability of LNG, demand elsewhere in the world, and prices for carbon and rivalling fuels like coal. The geography of the Asia-Pacific region (APR) means that the majority of traded gas is delivered as LNG, and in 2018 the region accounted for 74.9% of global LNG trade. Pipelines are starting to make inroads, but mainly from Central Asia – and now Russia – into China. In 2018, the APR only accounted for 9.2% of global pipeline gas trade.

On the basis of the above, we must differentiate between three large regional gas markets: a North American market, which is dominated by the United States and is largely self-contained; a European market that is increasingly import-dependent and supplied by pipeline from neighbouring producers, and also by a growing amount of LNG; and an Asia-Pacific market that is predominantly supplied by LNG, with China being both a significant producer of gas and an importer of both pipeline gas and LNG. As we shall see in chapter 4, the global LNG market is currently divided in two: an Atlantic basin and a Pacific basin, with a growing degree of interconnection.

Pricing Internationally Traded Gas
It should come as no surprise to learn that there is no global price for internationally traded gas (domestic gas pricing is equally complex, with most countries regulating the price of

gas to consumers). The oil market has different prices for different grades of oil and different benchmark prices, but it is possible to know the 'price of oil': in 2018, the average price for Brent Crude was $71.31.[29] Issues of price formation and contract structure are discussed in more detail in later chapters, but to keep things simple we will relate the different ways of pricing natural gas to the three-fold regional market structure described above.[30]

The two most important types of gas price formation for international trade in gas are: oil indexation (or oil price escalation) and gas-on-gas competition (hub-based, or market pricing). Originally, natural gas did not have much use and so there was no mechanism to price it. Once natural gas became more common as a feedstock, it was thus initially priced to a commodity that did have a price (mostly oil), and the gas price is determined through a base price and an indexation clause to competing fuels, typically crude oil, gas oil and or/ fuel oil. Depending on the oil price environment, oil indexation may result in a high gas price that makes it difficult to compete against alternative fuels – such as coal, in the case of power generation. In the case of gas-on-gas competition, the price is determined by the interplay of supply and demand and needs a mature and liberalized market environment (currently only North America, Australia and parts of the EU qualify). Gas is traded over a variety of time periods (daily, monthly, annually or longer) and trading can take place at physical hubs such as the Henry Hub – which is a physical distribution hub in Louisiana in the US – or via virtual trading hubs such as the National Balancing Point (NBP) in the UK.[31] Spot prices indicate what an amount of natural gas for immediate delivery costs, whereas futures inform us about how the market believes natural gas will be valued at a given point in time in the future. While most LNG is still oil-indexed, companies are increasingly experimenting with other ways to price LNG.[32]

Returning to the regional structure of the global gas market, gas trade in North America is based on gas-to-gas competition, and the key benchmark price is Henry Hub (though, increasingly, other hubs give a better reflection of local natural gas prices, e.g. Waha in the case of the Permian basin in West Texas, or Leidi Hub for the Appalachian). Canada has its own benchmark price, AECO, which stands for Alberta Energy Company. Because of the problems of market access mentioned above, this currently trades significantly lower than Henry Hub. In Europe, oil indexation long prevailed, but, starting in the 1990s in the United Kingdom, natural gas markets have slowly but surely been liberalized. Thus, continental Europe too has seen a trend towards gas-to-gas competition (the Dutch TTF – Title Transfer Facility – hub being the most liquid hub in the EU market at the moment). As markets opened, more traditional suppliers that have long favoured oil indexation have also moved in the direction of more market-based pricing, as illustrated by Gazprom's predominantly hybrid pricing system in 2019.[33] The future of gas trading and pricing in Europe is discussed in chapter 2. In the APR, the dominance of the LNG trade and a general lack of market liberalization means that oil indexation is still the dominant form of price formation, but the rise of LNG exports from North America and initial efforts to reform markets in countries such as Japan and China are increasingly putting pressure on this traditional way of doing business. This is discussed in chapter 4. Historically, there has been a degree of convergence between the different pricing systems, but since 2008–9 there has been a high degree of divergence. According to BP data, in 2017, the average Henry Hub price was 2.96 $/MMBtu; the average NBP price was 5.80 $/MMBtu; the average German import price (which includes some Russian oil-indexed gas) was 5.62 $/MMBtu; and the Japan LNG price was 8.10 $/MMBtu. As is explained in chapter 4, following the disaster at Fukushima, the shutdown of nuclear power production resulted in increased LNG demand in a tight market. The result was soaring LNG prices and a significant divergence between prices for LNG in Asia and for pipeline gas in the US and Europe. In 2018, the Henry Hub price averaged 3.13 $/MMBtu, the German import price 6.62 $/MMBtu and the Japan LNG price 10.05 $/MMBtu. Since then, though demand has continued to grow, significant new production has come onto the LNG market, and Asian and European prices have (at times) converged, but the US domestic gas price remains structurally lower, prompting the development of LNG export capacity. The impact that these trends are likely to have on the future price of gas – both the level of the price and the nature of price formation – and on contracting of natural gas are topics that we return to later.

A supply chain approach to gas security

Energy security is the subject of a vast literature and it is not our intention to add to it by engaging in a lengthy discussion of the meaning of gas security.[34] Using the IEA's definition of energy security, we can define gas security as: 'the uninterrupted availability of natural gas at an affordable price'. Chapter 5 discusses the complex interplay between energy security and climate change and the future role of natural gas, but for the moment we will focus on the narrower concern of gas security. There are two elements to this notion: first, 'physical security' or the continued flow of natural gas to meet demand at any given time; and second, 'price security', meaning the availability of that gas at an affordable price. Obviously, there is

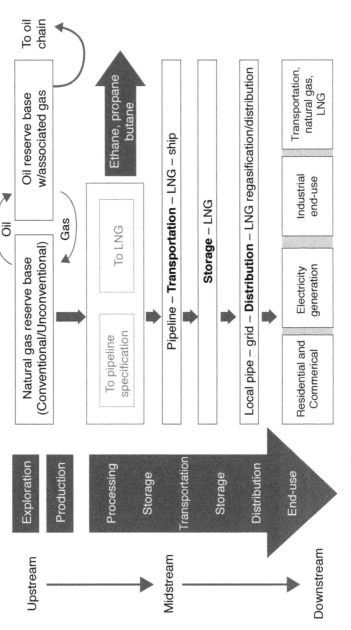

Figure 1.2 Natural gas supply chain

a trade-off between the two elements – for example, post-Fukushima Japan could only secure the LNG that it required by initially paying a high price. This section uses an explanation of the natural gas supply chain to unpack the elements that influence gas security in a particular national market.

Figure 1.2 presents a simplified version of the natural gas supply chain that enables us to relate the technical details discussed so far in this chapter to literature on the geopolitics of natural gas. The supply chain approach also highlights the importance of physical infrastructure and the role played by companies and markets in enabling the flow of natural gas from the wellhead to final consumer. As noted above, for the majority of natural gas production today, the supply chain remains within national boundaries; however, an increasing share sees the various elements being stretched geographically across continents and oceans, in no small part because demand growth almost entirely takes place in the developing world outside the OECD (Organisation for Economic Co-operation and Development).

Table 1.3 links the elements of the supply chain to the more detailed technical aspects that influence natural gas security. The upstream and issues of physical security of supply have attracted the greatest attention in relation to concerns about wider energy security. Gas-importing countries worry about their level of import dependence (how much of their natural gas consumption comes from imports) and their reliance upon particular suppliers of natural gas (diversity of supply). As mentioned above, while pipeline imports have tended to be less expensive than LNG, they do lock exporters and importers into an interdependent relationship. As we shall see in chapter 2, this is raising concerns in the EU where some member states are 100 per cent reliant on pipeline imports from Russia. The size of market share notwithstanding, various EU member states have

Table 1.3 A supply chain approach to gas security			
	Energy security	Dimensions	Issues
UPSTREAM	Security of supply	• Resource base • Technology • Investment	• Extent of proven reserves • Availability with existing technology and prevailing price • Access to reserves for investors • Access to investment to develop proven reserves
MIDSTREAM	Security of transport (transit)	• Processing • Transportation • Storage	• Processing of natural gas • Pipeline network • Compressor stations • Liquefaction facilities • LNG shipping • Regasification facilities • Storage • Interconnectors
DOWNSTREAM	Security of demand	• Power generation • Industrial use • Domestic use • Transport	• Role of gas in the energy mix • Price formation • Competitiveness • Contract structure • Energy policy • Carbon tax / Cap & Trade • Carbon capture & storage • Commercialization of cleaner 'gases'

demonstrated that building out infrastructure, and creating the option to import diverse supplies such as LNG, can function as a safeguard against market power abuse.[35] The earlier discussion about reserves suggests that, at the global scale, the physical availability of natural gas cannot be a major concern, though some traditional producing countries have exhausted their reserves and/or demand growth has outstripped production; more generally, we seem to be entering an age where midstream and downstream concerns are paramount. The one exception is the sustainability of gas production that, because of its nature, raises a different set of questions about security of supply (this is addressed in chapter 3).

The midstream involves all those elements of infrastructure

that are required to deliver gas to consumers. We have already discussed the distinction between pipeline gas and LNG, and both gas-exporting and gas-importing countries need to ensure that they have sufficient capacity to supply gas in a secure and affordable manner and in ways that limit the impact on the natural environment. Gas processing and transportation infrastructures are frequently the subject of environmental protest, though in most instances the impacts are modest once the construction phase has been completed. At the same time, transcontinental pipeline projects are multi-billion-dollar investments that require substantial political commitment and trust; they also raise the thorny issue of 'transit security', as clearly evidenced by the ongoing difficulties with transit of natural gas through Ukraine, which has been subject to political tinkering since 2006.[36]

Given the cost of infrastructure investment in the development of the upstream and midstream, a secure (and growing) market for natural gas is often essential to developing new sources of gas supply and routes of transportation. Natural gas has traditionally been used in three sectors: power generation, industry and households. Natural gas not only provides an energy service – heat – it is also an input into many industrial processes. In the power generation sector, it has to compete against other sources: oil, coal, nuclear power and renewable forms of electricity (primarily hydro, wind and solar). A modern Combined-Cycle Gas Turbine (CCGT) emits about half the amount of CO_2 per unit of electricity generated compared to a coal-fired power station. In addition, it can respond to changes in load (which happen more frequently in electricity systems with higher shares of intermittent renewables) better than conventional steam turbines, making it a possible partner of renewable energy – certainly as long as larger-scale storage of electricity is not commercially available. Gas power generation capacity is also far less costly and much quicker to build than a nuclear power station or a coal plant. Those who believe in natural gas as a bridging fuel see it replacing coal as a cleaner fuel with which to generate electricity, while cleaner options are built out simultaneously, as has happened to an extent in the United States and the United Kingdom.

Industry uses gas as both a source of heat and a raw material. Examples in which natural gas is used in the production process include the manufacture of steel (where natural gas is used to strip oxygen from iron ore), glass (where natural gas is combusted to heat furnaces to melt raw materials) and paper (to produce steam and power). Natural gas is used as a raw material mostly to produce fertilizers, hydrogen and ammonia. Natural gas liquids are also an important feedstock for the petrochemicals industry, where they are turned into, for instance, ethylene and propylene that are used to produce a wide variety of products such as tyres, plastics, sneakers and toothpaste. In the developed world, natural gas is widely used by households to heat water, warm homes and cook food. While a more recent trend has been to heat houses using electric heat pumps, it may still be the case that the electricity is produced from natural gas. Those who advocate a 'Golden Age for Gas' see not only the traditional uses of gas expanding (particularly power generation), they also see new sources of gas demand in the future – such as gas as a transportation fuel to replace more expensive and polluting oil products. The initiative by the International Maritime Organization (IMO) to curtail sulphur emissions from shipping, starting in 2020, has incentivized some shippers to opt for LNG as a cleaner fuel, even though competition from other cleaner fuels remains substantial, and uncertainties regarding the longer-term compatibility of LNG with anticipated additional environmental regulations are significant.[37] However, there remains considerable uncertainty over the ability of the natural gas industry to create the new demand necessary to warrant investment in the upstream and the midstream. As we shall see, climate change policy – in particular, placing a cost on emissions – has a key role to play here as it can raise the cost of higher-carbon alternatives, thus creating the opportunity for natural gas to fill a role as a bridging fuel.[38] However, gas-poor countries are not going to pursue policies that increase gas demand unless they are confident that they can access secure and affordable sources of supply.[39] The uncertainties over the future role of gas in the context of climate change policy are discussed in detail in chapter 5.

Towards a geopolitical economy of natural gas

It should be apparent from the discussion above that the contemporary natural gas industry is subject to a complex array of technical, economic, environmental and political forces. It is also clear that the material nature of natural gas as a fossil fuel influences the spatial organization of the industry; the majority of production is still consumed within producing states; long-distance, high-pressure pipelines often cross multiple borders when connecting producers and consumers; in the case of LNG, a complex spatially disaggregated supply chain enables stranded gas to reach markets beyond the reach of pipelines and is increasingly adding consumers to create a more globally interconnected market. The three dimensions of the geopolitical economy approach adopted in this volume provide both an analytical framework for considering the key dimensions of the industry and also an integrating framework for considering the interrelationship between these dimensions. Geography – the friction of distance – imposes costs and determines just how far production from one region can travel to another. Because of the role that pipelines play and the fact this trade creates 'interdependencies' between states, the state is often implicated, to various degrees, in natural gas development. As we shall see, the geographical and the political dimensions come together to create a 'geopolitics of natural gas' that is critical to understanding the industry today, and impacts investment decisions, and sometimes even trade flows. That said, economic fundamentals related to the cost of production, processing, transportation and consumption also impose the logic of the market on the global trade flows of natural gas. For example, the US Department of Energy's declaration that LNG exports represent 'freedom gas' that can reduce Europe's dependence on Russian pipeline gas, is political rhetoric: LNG will only flow to European import terminals if the companies involved in that trade can make a satisfactory profit.[40] Nevertheless, some states – Poland, for example – may choose to import more expensive LNG to reduce their reliance on Russian pipeline gas, demonstrating that physical security comes at a cost. The three chapters that follow deploy a geopolitical economy approach to explore the critical factors shaping the current and future role of natural gas in the global energy system. We start with an analysis of the more traditional way of trading natural gas, pipeline trade. Subsequently, we examine two fundamental recent changes that have shattered conventional wisdoms in natural gas markets: the shale revolution, and the rise and democratization of LNG. The analysis then turns to the future role of natural gas in the context of governments' attempts to curtail GHG emissions in global energy systems, and price pollution. This chapter discusses the underlying tensions between these trends of supply, trade and decarbonization, and opportunities for their reconciliation. The concluding chapter returns to the issues raised in this introduction and considers the factors shaping the geopolitics of natural gas.

CHAPTER TWO

Pipeline Geopolitics

As we noted in the previous chapter, the majority of the world's natural gas production is not traded across international borders. Of the 32% that is traded internationally, the majority (65.1% in 2018) is transported via pipeline, but over the last decade the relative share of pipelines in global natural gas trade has declined as LNG trade has grown at a faster rate. Some analysts have suggested that global LNG trade will surpass pipeline trade around 2035 – a fundamental shift with major consequences that we will discuss later.[1] Back in 2008, pipelines accounted for 72.2% of total traded gas. Despite its falling relative share, the volume of gas pipelines has grown significantly from 587.3 bcm in 2008 to 805.4 bcm in 2018, illustrating overall demand growth. Both then, and now, pipeline gas trade is concentrated in Europe (including Turkey and Ukraine), which in 2018 accounted for 59.5% of international pipeline gas trade.[2] At the heart of this trade pattern are two significant reserve-holding states: Russia (19.8% of global gas reserves) and Turkmenistan (9.9% of global gas reserves).[3] As we shall see, historically, flows of pipeline gas from Russia and Central Asia (and, prior to 1991, the Soviet Union) have dominated concerns about pipeline geopolitics. In a European context, Norway is an important supplier of pipeline gas, but its close relationship with the EU has meant that it seldom features in discussions of pipeline geopolitics. In southern EU member states, natural gas trade with Algeria and

Libya is subject to geopolitical debate, albeit with a substantially less explicit media profile than EU–Russia trade. More recently, the emergence of China as a market for pipeline gas is creating a new centre of attention. Thus, this chapter focuses on pipeline geopolitics across the Eurasian landmass. Clearly, there is international pipeline trade elsewhere, such as Latin America, North America, and the Middle East and North Africa (MENA) region, but it is developments in Eurasia that have been of global significance and that are synonymous today with the notion of 'pipeline geopolitics'.

The Eurasian continent has substantial natural gas reserves, totalling 32.5% of the global total (see table 2.1). Production and trade are dominated by two countries, Russia and Turkmenistan; but their producing fields are thousands of kilometres from major consuming regions in Russia and the rest of Europe. Long-distance, transcontinental, large-diameter pipelines represent the most economical way of transporting this energy carrier. Most recently, gas has started to travel eastwards as a new transcontinental network is emerging to

Table 2.1 Eurasian gas: reserves, production, consumption and trade					
	Reserves at end of 2018 (tcm)	Share of global total reserves at end of 2018 (%)	R/P years	Production 2018 (bcm)	Consumption 2018 (bcm)	Trade* 2018 (bcm)
Azerbaijan	2.1	1.1	113.6	18.8	10.8	9.2
Kazakhstan	1.0	0.5	40.7	24.4	19.4	25.6
Russia	38.9	19.8	58.2	669.5	454.5	223.0
Turkmenistan	19.5	9.9	316.8	61.5	28.4	35.2
Ukraine	1.1	0.6	54.9	19.9	30.6	–
Uzbekistan	1.2	0.6	21.4	56.6	42.6	14.0

* pipeline trade only

Source: BP (2019), BP Statistical Review of World Energy, June 2019. London: BP, pp. 30–41.

satisfy China's new thirst for gas. This chapter is structured as follows: the next section considers what it is about international gas pipelines that presents geopolitical challenges. The third section then turns to the specific case of gas pipeline trade between Russia and Europe and highlights how the changing political geography of Europe and the growing disconnect between the worldviews of Brussels and Moscow have complicated mutual relations. The fourth section turns to Central Asia where the collapse of the Soviet Union stranded the region's significant gas reserves, landlocked at the heart of the Eurasian continent. When Moscow and Gazprom eventually turned their backs on Central Asia, Chinese state-owned enterprises stepped in as new buyers of pipeline gas; however, Turkmenistan, in particular, found itself swapping one dependency for another and is now exploring other alternatives. The fifth section explores Russia's strategy to reduce its dependency on European markets by developing new pipeline infrastructure to supply, amongst others, China. Ironically, Russia's so-called 'Eastern Strategy' or 'Asian Pivot' finds itself competing with its former partners in Central Asia to support China's natural gas ambitions. The chapter concludes by returning to the fundamentals of 'pipeline geopolitics' to determine what lessons can be learnt from the Eurasian case.

The fundamentals of pipeline geopolitics

What is it about international natural gas pipelines that makes them vulnerable to geopolitical manipulation? As we shall see in this chapter, the current configuration of pipelines across the Eurasian continent has evolved over the last fifty years, but at its heart are a number of very large gas fields – most significantly in the West Siberian basin in Russia – that have been connected to large gas markets, in both European Russia and what was once known as Eastern and Western Europe. As the next section explains, it is changes in the underlying political geography that have complicated the situation since 1989.[4] The basics of an international pipeline system are, on the face of it, quite straightforward and relate to the supply chain that was explained in the previous chapter. At one end (the upstream), it is necessary to have a large gas reserve that can provide a secure supply of natural gas over a period of decades; at the other end (the downstream), it is necessary to have a stable level of demand, with buyers that can afford to pay a price that more than covers the cost of production, processing, transportation and storage (though various parts of the supply chain can be subsidized for political and social reasons). Between the two, it is necessary to construct a pipeline system. This tends to evolve incrementally as gas demand grows, but, as we shall see, in recent years new pipelines have been built for not exclusively economic reasons.

In geopolitical terms, this simple system of supply, transport and demand can be translated into three dimensions of energy security, all of which complicate matters. First – and the issue that has attracted greatest attention in Europe – is the notion of 'security of supply', which is focused on the secure and affordable physical supply of natural gas to consumers. This is the primary concern of gas-importing states, who, once locked into a dependence on imports of pipeline gas, are vulnerable to disruption motivated by economic and/or geopolitical gain. Second is the notion of 'security of demand', which relates to the need for the owners of the upstream reserves and investors in midstream capacity to be certain that there is sufficient long-term demand and an ability to pay a price that can justify the development of those reserves and any associated infrastructure costs (processing plants, pipelines and compressors). Third is the notion of 'transit security', which relates to transnational pipelines, which, by definition, cross one national boundary or more and are thus reliant on transit states to allow the initial construction of the pipeline and then the safe passage of gas, and the levy of predictable and competitive transit fees. It is often the case that the transit state is also a consumer of the exported natural gas.

Two further factors need to be considered when it comes to pipeline geopolitics, and they stem from the fact that an international pipeline represents a very significant upfront capital investment that becomes a sunk cost once paid for – but it is a large immovable infrastructure that ties the exporter into a long-term relationship with importers. First, as noted above, from the viewpoint of the importing state (or company), there is a desire to avoid becoming too dependent on one source of supply, and the notion of 'import dependency' is often used to describe this issue. The antidote is to pursue a policy of 'diversification', in terms of sources of gas supply, the routes to market and the level of reliance on gas in the energy mix. Second, the same concerns also exist for the exporting state, which does not want to be overly reliant on a single export market. Equally, the exporting state requires security of demand and is often dependent upon export revenues to cover its costs and meet the state's budgetary requirements – thus, there is actually an 'interdependence' between exporting and importing states.[5] Nonetheless, the fixed nature of the pipeline does mean that it is vulnerable to interruption at any stage from production to final consumption, but most often a supply disruption is a consequence of technical failure, rather than geopolitical intrigue. These fundamentals will become clearer as we investigate the pipeline geopolitics that have emerged as a consequence of the development of transcontinental pipeline systems across the Eurasian continent.

Europe–Russia natural gas trade

Contemporary discussions about natural gas trade between European buyers and Russian sellers at times implicitly suggest a grand scheme developed by Russia's state-controlled gas company Gazprom to – in the words of US President Trump – 'totally control' Germany and the larger part of the EU.[6] Reality could hardly be further from the truth. Major natural gas discoveries in the Soviet Union date back to the Second World War, when Stalin was in search of oil to fuel the nation's defence against Nazi Germany. In that search, major finds of natural gas were not uncommon, but were generally considered useless because there was no market for the product, a response that would last well into the 1970s in some parts of the world. It was not until well after the war, with the Soviet Union in recovery mode, that natural gas increasingly was considered a useful fuel. Austria, which was occupied by the Soviet Union right after the war, was amongst the first countries to develop its domestic gas market. It did so not because the Soviet Union offered the commodity, but because local companies had successfully developed both oil and gas in the inter-war period. After the war, natural gas became an increasingly popular fuel in industry, and for space heating and power generation. Something similar happened in the Netherlands, incentivized by the discovery of the massive Slochteren field in 1959. In Austria, however, domestic reserves proved unable to keep up with growing demand, and thus Austrian companies – often in competition with one another – started to develop plans to bring additional resources from abroad.[7] Connecting the country to the gas grids in Czechoslovakia to the east proved to be the winning strategy, and today Austria has a mature gas market.

Several of the key markets for Gazprom, which has a

monopoly on gas exports from Russia by pipeline (more about this later), have similar histories when it comes to the development of natural gas markets in their countries – namely, an organic process of increased trade between (at the time) publicly owned companies. It is worth noting that, as natural gas trade in the then European Economic Community grew, the fuel became a central part of post-1973 attempts by OECD countries to become less dependent on imported oil and exposed to the actions of OPEC. Ironically, the first concerns about growing dependence of West European countries on Soviet natural gas came from across the Atlantic Ocean, where in the early 1980s US President Reagan warned European allies that becoming overly dependent on the Soviets was not in their interest (also because revenues from gas trade were considered a lifeline for the Soviet regime, and its military).[8] This was one of the few occasions when the UK's Prime Minister Margaret Thatcher failed to see eye-to-eye with President Reagan. European leaders essentially dismissed the American concerns, and natural gas trade went back to sleep mode, from a geopolitical point of view. Of course, as we shall see later, these events have contemporary resonance as the Trump Administration has expressed concern about Europe's continued and growing reliance on Russian gas.

By the late 1980s and early 1990s, the seeds were sown for this status quo to be fundamentally altered. First, in 1989, the Berlin Wall fell. This had major consequences for the configuration of Europe's political map, and, later, for the shape and functioning of the EU, which was founded in 1993. Second, in the early 1990s, policymakers in Brussels and those in various European capital cities, with the UK leading the way, embarked on a journey of market liberalization. The idea, essentially, was that markets that had historically been orchestrated centrally and by state-owned enterprises, had to be liberalized, to compete with one another for the ultimate benefit of the consumer. Governments, the belief was (and still is, though the EU is at a turning point – more in chapter 5), set the framework within which market actors invest and compete, under supervision of (preferably independent) regulatory authorities. Infrastructure was to be separated from natural gas production and trade – so-called unbundling – and considered a public good that all citizens should have access to. Various key markets in the EU became subject to sweeping market reform, and those for natural gas and electricity were amongst them as Brussels sought to create a single internal energy market for gas and electricity.

These developments turned out to be important for European natural gas markets in a number of ways. First, they resulted in an EU that has developed in different places at different speeds, in part by design, and in part because the historical context forced it to. Market reform has chiefly been orchestrated from Brussels by issuing legislative guidelines for the member states to implement. Most of these guidelines were given in the form of directives; some, in the form of regulations. This is relevant because regulations translate directly into national legislation, while guidelines give member states a substantial amount of liberty regarding their implementation: they basically prescribe an end-goal, not a means to get there. Logically, some member states were more forward-leaning than others, with the United Kingdom typically leading the way, and member states such as Germany and France being more protectionist of their domestic champions. It is important to add that current EU member states in Central and Eastern Europe had not joined the EU by this time: they did so only in 2004 (the Baltic states, Poland, Czech Republic, Slovakia, Slovenia, Hungary, Malta and Cyprus), 2007 (Romania and Bulgaria) and 2013 (Croatia). Therefore, these member states were exposed to the idea of market liberalization significantly later than their EU

Table 2.2 Russia–EU gas trade	
	Imports from Russia (mcm), 2018
Austria	13,091
Belgium	–
Bulgaria	2,874
Croatia	1,225
Cyprus	–
Czech Republic	8,343
Denmark	1,720
Estonia	396
Finland	2,383
France	9,047
Germany	54,788
Greece	3,085
Hungary	7,642
Ireland	–
Italy	28,103
Latvia	1,384
Lithuania	1,289
Luxembourg	–
Malta	–
Netherlands	8,093
Poland	9,303
Portugal	–
Romania	1,725
Slovakia	7,247
Slovenia	–
Spain	–
Sweden	–
United Kingdom	4,400

Source: authors' compilation based on various sources: *BP Statistical Review of World Energy*, June 2019; Gazprom 2018 delivery statistics; IEA monthly gas flow data 2018; Eclipse (S & P Global Platts Analytics).

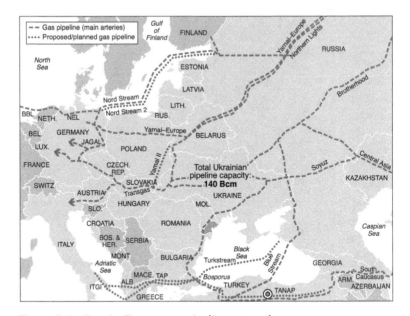

Figure 2.1   Russia–Europe gas pipeline network

Source: reproduced with permission from S & P Global Platts, ©2018 by S&P Global Inc.

peers, and, arguably, to this date most are behind the curve. By this, we mean that in these countries the idea of market liberalization, and promoting competition, has not had as much time to take root as in Northwestern Europe. Consequently, in several of these countries features like price regulation are still common, and competition is lacking (and sometimes viewed with suspicion). This is important as some of these countries are also the most exposed in terms of reliance on Russian gas imports (table 2.2).

Second, market liberalization slowly but surely created a mismatch between market actors on the one hand, and governmental agendas on the other. Once liberalization took root, companies felt the direct impact of the newly created single

European market, for better or for worse, whereas legislatures retained their national focus, a problem that persists to this day. To illustrate the former, the integrated gas companies that lost important assets (their networks, which gave them exclusive access to a client base) were soon at risk of losing relevance, and some went under. In political discussions, however, the focus continued to be on the nation state – as exemplified, for instance, by the continuing concern of the European Commission that too often problems in the gas sector are not dealt with collectively, but, rather, unilaterally.[9] For quite some time, regulatory authorities maintained their national orientation, even though, with the establishment of the Agency for the Cooperation of Energy Regulators (ACER) in 2011, an official European institution has existed to monitor the functioning of the internal gas market.

Third, in retrospect, we can observe that the EU and Russia, its key supplier of natural gas, slowly but surely started drifting apart. There was likely no way of knowing this at the time, and one working hypothesis (advocated by the German government, for instance) had been that, by keeping Russia close, it might be tempted to embrace market liberalism. The opposite happened.[10] The Kremlin – and, by default, Gazprom – started signalling that they had no interest in joining the European gas market project and did not intend to unbundle Gazprom. It is worth noting, though, that, domestically, Gazprom has been forced to allow third-party access to its networks (and lost a significant part of its market share to companies like Novatek and Rosneft), and that it started operating an electronic sales platform in the EU, and offering products at the St Petersburg International Mercantile Exchange, SPIMEX. To this day, lawsuits continue about the European single market project, and the position and behaviour of Gazprom herein. In 2015, European Commissioner Vestager charged Gazprom with abusing its dominant market position in Central and Eastern Europe, and possibly limiting customers' ability to resell natural gas, and charging unfair prices. In 2018, the European Commission then imposed binding obligations on Gazprom to remove all remaining contractual barriers to the free flow of natural gas, facilitate the free flow of natural gas to and from isolated markets, ensure competitive prices, and cease to leverage its dominant role – e.g. in infrastructure.[11]

Fourth, the eventual expansion of the EU had major implications for the way in which European institutions thought about a resource like natural gas. In Northwestern Europe, where markets are now well functioning, support for the single market persists, and why would it not? In parts of Central and Eastern Europe, however, natural gas, even though consumed only in modest volumes, is viewed with suspicion, not least because most of it comes from Russia, a country that most of these member states have a troublesome history with. Some of them have even suffered through gas supply disruptions, so why would these countries not have mixed feelings?

By the mid-2000s, European gas market development was reasonably well under way. In Northwestern Europe, the United Kingdom had shown the way by privatizing (this started in 1986) and unbundling (completed in 1997) its integrated gas company, and, in the early 2000s, other member states followed suit, such as the Netherlands with the dismantling of the Gasgebouw into GTS and trading company GasTerra. Companies and transmission system operators were investing in interconnectors between various countries, gas from various sources (the UK, Norway, Netherlands, Russia, LNG) could increasingly flow better to meet local demand, and prices in this part of the continent were converging. The internal market had started to work, when Russia's Gazprom and the Ukrainian oil and gas company Naftogaz got into a dispute in the spring of 2005, which escalated in winter of that year, and, on 1 January 2006, Gazprom shut off supplies through Ukraine, only to re-establish them on 4 January after a preliminary agreement between both sides had been reached. In the winter of 2009, a new dispute between Gazprom and Naftogaz caused a more serious supply disruption, which lasted thirteen days and forced several member states to switch to alternative fuels and reduce industrial activity.[12]

The 2006 incident, despite its modest impact, set the tone for the years to come. Gazprom's response to the dispute with Naftogaz was, in cooperation with several Northwest European companies, to construct gas transport infrastructure bypassing Ukraine, in order to reach clients in Northwestern Europe directly, and limit the transit risk, continuing the trend that had already been set by the construction of Yamal–Europe through Belarus to Poland, and Blue Stream to Turkey. As such, the disputes in 2006, and later 2009, convinced the Russian leadership that transit risk reduction had to be accelerated. Nord Stream 1, as the proposed pipeline was called, was controversial from the onset. It had support in Northwestern Europe, where the January supply disruption was mostly blamed on Naftogaz, but was viewed with great suspicion by EU member states to its east and southeast. The then Polish Defence Minister Radek Sikorski went as far as calling the proposed project the Molotov–Ribbentrop pipeline, referencing the deal between Stalin and Hitler in 1939 to partition Poland.[13] Next to geopolitical fears, there were also environmental concerns, mostly voiced by the Scandinavian member states, about building this pipeline through the Baltic Sea from Russia to Germany. Eventually though, all the necessary permits were approved, the project secured funding, and the first natural gas flowed in 2011, cheered by President Putin and a handful of European heads of state. Concerns about the pipeline did not materialize in the first years of operation, and once pending midstream and regulatory issues in Germany were resolved, the pipeline operated almost at full capacity starting in 2017.[14]

Concerns about single-source dependence, voiced mostly by member states in Central and Eastern Europe, did not pass unnoticed though, and continued to dominate policymaking in Brussels. The antidote that European institutions, particularly the European Commission, had been prescribing was further market integration and cooperation, amongst other things.[15] By opening up markets for competition and building infrastructure to allow the free flow of natural gas, the chances of market power abuse would by definition be reduced. It worked in Northwestern Europe and could yet work in the entire EU. However, history had already shown that this was easier said than done. Two fundamental challenges apply to most member states in Central and Eastern Europe. One, gas consumption in these member states is typically modest, as most of them (with the exception of Romania) do not have significant domestic resources and have developed their economies using other feedstocks, mostly coal and oil. As a result, investing in infrastructure – albeit pipelines, interconnectors or reverse-flow facilities on existing infrastructure – is not commercially attractive for most investors. In these instances – of what economists call a 'market failure' – intervention from public actors can help nudge the market in the desired direction. But there was a second challenge: public actors in these member states do not have the same mindsets as their peers in Northwestern Europe, or Brussels. After all, until 1989 many had been part of a totally different economic system and they had a ten- to fifteen-year time lag in terms of reforming their markets and opening them up for competition. For example, even though Poland has built interconnectors with neighbouring countries such as Germany, it has also made a significant investment in regasification capacity to import LNG, and its national oil and gas company PGNiG (Polskie Górnictwo Naftowe i Gazownictwo: Polish State Oil and Gas Company) signed a twenty-year contract with Qatargas to bring LNG into Poland at a price that was higher than what the company used to pay Gazprom. The initial agreement – based on oil indexation – was signed in 2009 and deliveries started in 2015; PGNiG has since doubled the size of its agreement with Qatargas.[16] This shows that the company, and Polish authorities – which historically have been very closely intertwined – do not subscribe to the fundamental premise of EU gas market integration: that competition is the way to safeguard security of supply, not having your citizens pay over the odds. Another general misconception is that member states in Central and Eastern Europe all have similar views regarding the risks that come with importing natural gas from Russia. As usual, reality is more nuanced, with, for example, the Czech Republic endorsing the Brussels mantra of market integration and strong regulatory safeguards, whereas other member states such as Poland and Hungary are more sceptical of this.[17]

In their efforts to develop their gas markets, member states from Central and Eastern Europe have tended to look at Brussels for support, and understandably so. Before the new members joined the EU, Brussels' policymakers had often suggested that the newcomers should clean up their energy economy, both to improve local air quality, and to reduce (at the time an emerging political concern) GHG emissions. These member states in turn were willing to do so but counted on Brussels for (financial) support. Once in the EU, this (mostly) proved to be wishful thinking as Brussels has a very modest mandate when it comes to most aspects of energy policy. Member states in the west of the continent often pointed to their companies who had made their own investments to develop gas markets and

wondered why citizens throughout the EU had to foot the bill
to develop gas markets in Central and Eastern Europe. Today,
commercial infrastructure in the EU seems to have stopped
being built, and most new infrastructure projects, such as interconnectors, reverse flows or new pipelines, require significant
public support – for instance, through the so-called Projects
of Common Infrastructure, a Brussels subsidy scheme.[18] Thus,
gas market development and integration in Central and Eastern
Europe have continued to be problematic, even though since
2011 meaningful progress has been made.[19] As noted earlier,
European leaders agreed to establish ACER in 2009 (it started
operations in 2011) to help synchronize gas market rules, and
the European Commission was given a modest mandate to help
orchestrate infrastructure projects in cases of market failure.
Its overall budget for these projects continues to be modest,
however (< 6 billion euros for the period from 2014 to 2020
under the Connecting Europe Facility), which, realistically,
mostly allows it to fund feasibility studies. A rare exception
is the interconnector between Lithuania and Poland, which is
partly funded by the Commission, after intervention by ACER
in relation to how to divide the costs between the involved
member states (this had been the chief stumbling block of the
project for several years).[20] On the up side, one can observe real
progress in market integration in the EU, yet it may take years
to complete the internal market, despite former EC President
Barroso's repeated claims that the project would be achieved
by 2014, and similar claims in 2019 from his successors.[21] In
addition, not all answers can come from Brussels. Analysts
must continue to ask themselves: if dependence on Gazprom is
a security threat, as often voiced by representatives from various European countries, why have local politicians, regulatory
authorities or business representatives done so little to improve
access to alternative sources of supply, and to foster competi-

tion? And what does this discrepancy between alleged threat and tangible action tell us about the real nature of this threat? As convenient as it might be to point the finger at Moscow, leadership will have to come from cities like Warsaw, Sofia and Budapest. As the example of the interconnector between Poland and Lithuania shows, it is often reasons in the sphere of the local political economy that prohibit projects from moving forward, rather than Russian intervention.

In 2014, EU–Russian relations hit a new low with the annexation of Crimea, and the military intervention of Russian troops into Eastern Ukraine, in support of a local insurgency. The frozen conflict that emerged has lasted, and there is currently no indication to suggest a solution will be found soon. The war in Ukraine incentivized the European Commission to perform so-called 'stress-tests': running various scenarios of gas supply disruptions, to see whether European member states would be able to withstand such an event.[22] The outcomes, in short, generally confirmed that European efforts to develop and further integrate gas markets had been successful, and most member states were – admittedly by taking at times draconian measures, and assuming full cooperation with each other – able to keep their energy systems up and running in case of a full supply disruption. This was good news in bad times, but the lesson that most policymakers seemed to take from it was rather different: dependence on Gazprom had to be further reduced.[23] Meanwhile, Russia and Gazprom drew their own conclusions from the conflict – namely, that Ukraine as a transit state had lost its credibility, and that a set of additional pipelines (Nord Stream 2, from Russia to Germany, and Turkstream, from Russia to Turkey and/or Southeastern Europe) had to be built to circumvent the country once and for all (an idea that had been supported starting in 2006). In other words, Gazprom adopted a strategy to reduce or possibly even eliminate transit risk by proposing various projects to circumvent Ukraine. Ironically, Ukraine in turn also addressed its reliance on Russian pipeline gas imports, by ceasing direct imports from Gazprom, and instead re-importing Russian gas from its European neighbours (mostly Slovakia). This, in combination with changes in transit fees, further undermined the economics of the Ukrainian natural gas transit, ahead of the expiration of the existing transit agreement in late 2019.[24] An agreement was finally reached on 30 December for the period 2020–4, but, longer term, the issue of natural gas transit through Ukraine will remain uncertain as Russia has other options via Nord Stream and Turkstream.

Nord Stream 2 has proved to be an issue that the European member states were not able to address collectively. Instead, the pipeline has created clear dividing lines between several member states. Opponents pointed to fundamental problems such as a growing market share for Gazprom (Russian pipeline imports accounted for 34.1 per cent of the natural gas consumed in the EU in 2017, and 36.7 per cent in 2018) and voiced more practical issues like loss of transit revenue for national companies, or environmental concerns. Proponents point to the low-cost gas that Gazprom offers, and the fact that (a part of) European industry is required to compete globally, and that consumers generally want to pay as little as possible. Initially, the European Commission attempted to be an honest and neutral broker, but soon it picked sides and indicated that the pipeline would run counter to its ambition to diversify European gas supplies. The fundamental challenge, though, has been that policymakers do not buy natural gas, and at the end of the day this is a decision that companies (in this case, state-owned Gazprom plus five financial supporters from four EU member states) have to make, if they can obtain the relevant permits and secure funding. There is an obvious contradiction between the European Commission seeking to liberalize and create a single European gas market, and then wishing to intervene when market actors seek solutions that do not suit its political agenda. Somewhat ironically, various legal actions brought by the European Commission and individual member states, together with market forces, have forced Gazprom to embrace a gradual move away from oil indexation to hub-based pricing, though a significant amount of its sales to Europe remain in legacy long-term contracts that will not end until well into the 2020s.[25] Thus, Gazprom's position of market prominence will continue, despite EU concerns and US criticism.

The controversial nature of Russia's gas supplies to the EU were further demonstrated when the US government started to lobby actively against the Nord Stream 2 project. This effort started under the Obama Administration, and officials repeatedly voiced their concerns about the growing dependence of their European allies on Russian natural gas.[26] Under the Trump Administration, following a series of official statements, the impression was raised that the key American concern is in fact to ensure that American companies can sell LNG in the European market.[27] Thereafter, proponents of Nord Stream 2 called out US concerns as mostly self-serving. The reality is surely more nuanced, and the merits and pitfalls of US energy diplomacy deserve more empirical attention.[28] At the time of writing, the Nord Stream 2 company had started laying pipe in both Germany and Russia, and the chances are high that construction will be completed. The more important overarching consequence of this saga, however, has been a further deterioration of the image of natural gas as a fuel source in the EU. This image has further been challenged following concerns over induced seismicity in the Netherlands – until recently the EU's largest producer of natural gas – and growing dissatisfaction amongst NGOs and a part of the population about the pace of the energy transition, in which a future role for natural gas is by no means guaranteed.

Hence, the outlook for natural gas in the EU is uncertain and one of the unintended consequences of Russia's strategy to circumvent Ukraine may be to constrain future European gas demand. That said, there is a lot of buzz about biomethane, power-to-gas technologies, and hydrogen as part of a 'natural gas 2.0' that would be compatible with a low carbon agenda, though many uncertainties remain about the long-term commercial potential of some of the options being debated. We discuss this in more detail in chapter 5. Meanwhile, political discussions about EU import dependence continue on both sides of the Atlantic Ocean, but the market seems to have made up its mind and import levels of Russian gas into the EU have risen to record highs as gas demand has recovered with the economic recovery starting in 2013/14.[29] With modest appetite for domestic exploration (see the later discussion of shale gas), Russian natural gas is an inevitable part of the future European fuel mix. Whether this will all be supplied by Gazprom is less certain, as exports of LNG from Russia by other companies (e.g., Novatek) have been allowed since 2013 and started in the spring of 2018. Additional turmoil related to Gazprom may incentivize the Kremlin further to liberalize exports from Russia in general, though the chances of this seem slim at the moment. It is also uncertain whether the European Commission will intervene in the market functioning that it has advocated for so long, now that the outcome of that liberalized market results, unsurprisingly, in a significant market share for Russian gas. We call this development 'unsurprising' because several alternative sources of supply have not materialized as hoped: the Southern Gas Corridor will only bring modest volumes of natural gas into southern Europe, shale gas production in the EU has not materialized and has little chance of doing so (as we discuss in detail in chapter 3), increased imports from North Africa continue to be problematic, and, most recently, Dutch production has dramatically been curtailed.

The post-war history of Europe's gas market – particularly events since 1989 – highlights the importance of a geopolitical economy approach to understanding the geopolitics of natural gas. The majority of the underlying transcontinental pipeline infrastructure was built in the 1970s and 1980s, largely under state ownership and control. However, since 1989, all three aspects of the geopolitical economy of Europe's gas market have fundamentally changed. First, the political geography of the region has been transformed following the collapse of the Soviet Union, new independent states have emerged in former Soviet space, and most countries in Central and Eastern Europe have joined the EU. Consequently, new transit states have emerged between Russia and European consumers, and sensitivities regarding import dependence on Russia, which were arguably not there before, have entered EU policy circles. Second, the political setting for Russian gas exports to Europe has also changed. After initial hopes that Russia would pursue liberal market reforms, it has become clear that there is now a fundamental difference between the economic model and worldview in Moscow and that in Brussels and the EU's member states, although there are also varying views within the EU. Moscow has embraced the zero-sum world familiar to realist theories of energy geopolitics with their belief in the primacy of state actions, while the EU has embraced the market-based approach of (neo)liberalism with its belief in market principles.[30] However, as this analysis has shown, the full creation of a single European gas market has yet to be realized. Although Northwestern Europe now benefits from a fully integrated liquid gas market, problems remain elsewhere in the EU. Third – and the final irony – is that, because Russia represents the lowest-cost source of gas imports for most of the EU (for some member states, this is pipeline gas from Algeria instead), a reliance on market principles has increased Russia's market share despite the difficult political situation between Russia and the EU as a result of the ongoing conflict in Ukraine. Furthermore, political intervention to promote diversification may lead to changes in the EU's natural gas mix – something that has happened in Poland, where policymakers do not want to rely on market forces to safeguard security of supply, a mantra that has been leading in Brussels for several decades, but possibly has run its longest course there too.

Central Asia: between a rock and a hard place

Any discussion of the geopolitics of Central Asian gas must start with an appreciation of the region's location at the heart of the Eurasian landmass (see figure 2.2). The region possesses significant natural gas resources, but the fact that they are land-locked and some distance from major centres of consumption means that exporting states – principally Turkmenistan – are reliant on the cooperation of transit states, some of whom have their own gas resources that they are seeking to monetize – Russia and Iran, to be specific.[31] The following discussion focuses on the experience of Turkmenistan at a time when it has swapped dependence on Russia for dependence on China.

The exploitation of Central Asia's gas reserves started in the 1960s and the subsequent development of pipeline infrastructure was guided by decisions made in Moscow that served the wider needs of the Soviet Union. Construction on the Central-Asia-Centre (CAC) pipeline started in 1965, and the pipeline was inaugurated in 1967. Initially, it linked the Gazli field in

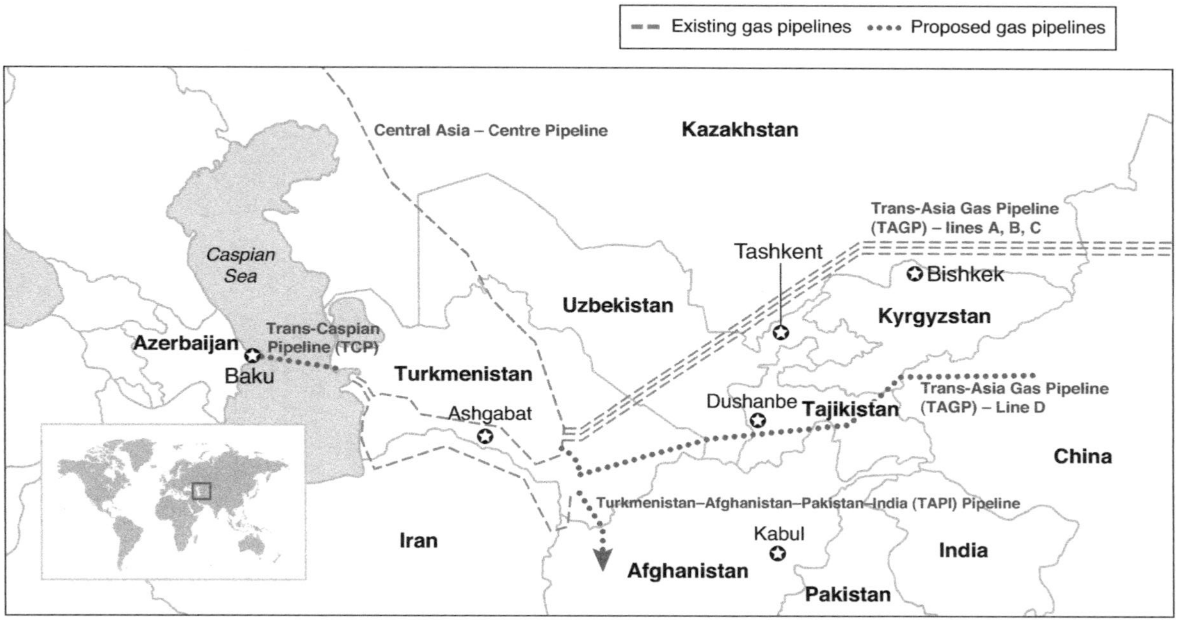

Figure 2.2 Central Asian gas pipeline network

Source: reproduced and modified with permission from Stefan Hedlund (2019), *Turkmenistan Comes into Focus.* Vaduz: Geopolitical Intelligence S.ervices AG. Available at www.gisreportsonline.com/turkmenistan-comes-into-focus,politics,2834,report.html.

Uzbekistan to industrial regions in European Russia. Gradually, the pipeline network expanded, in five stages, to collect additional gas from Turkmenistan, Kazakhstan and Uzbekistan. By the end of 1980, the CAC had a total annual capacity of 90 bcm.[32] This flow of natural gas to Russia was part of a form of 'internal colonialism' within the Soviet Union, whereby Central Asian republics supplied the European republics with raw materials and agricultural products, in return for financial flows that supported the development of the region. However, it locked them all into a dependent relationship that left them ill prepared for the collapse of the Soviet Union in 1991.

When the Soviet Union collapsed, the newly independent Central Asia states had little alternative but to continue to supply Russia with their natural gas. The CAC Pipeline came under the monopoly ownership of Gazprom and served Russian interests. Barkanov suggests that Russia has sought control over Turkmenistan for two reasons: first, to provide an alternative source of supply to meet domestic demand in Russia; and second, to prevent competition with Gazprom in the lucrative EU market. Further, he maintains that Russia faced a choice in how it handled this relationship, between what he calls 'winner takes all' and 'rent-sharing'.[33] During the 1990s, Gazprom paid Turkmenistan a low price akin to that paid by domestic consumers in Russia and then sold that gas, either directly or indirectly, to its European customers for two or three times that purchase price. It also often unilaterally decreased the volume of gas imports for both economic and political reasons. Furthermore, prior to 2000, only 40 per cent of payments were made in cash, with the balance being in the form of barter for food and commodities.[34] Clearly, Russia and Gazprom pursued a policy of 'winner takes all', aided and abetted by the first Turkmen President Saparmurat Niyazov Turkmenbashi, who was aligned with Moscow. During this time, Gazprom saw little need to invest in the CAC and the utilization rate of the pipeline network started to fall – between 1991 and 2008, it carried an average of 35–40 bcm, less than half its nameplate capacity.

When Vladimir Putin came to power in 2000, he sought to improve relations with Central Asia to reaffirm Moscow's influence over what it described as the 'near abroad'. This was prompted by growing US and European interest in the region's energy resources. In addition, rising gas prices in Europe made it more palatable to share the rents with Central Asia. This resulted in a gradual improvement in the terms of trade, as well as a commitment to renovate the CAC, but falling volumes – Russia had agreed to take 80 bcm, but never did – were not adequately compensated for by higher prices. Just as it became clear that Russia was failing to deliver on its promises, a new suitor for Central Asian gas emerged, in the form of China.

Then, in 2006–9, a complex set of circumstances dramatically changed the geopolitics of Central Asian gas exports. First, in 2006, President Turkmenbashi died and was replaced in February 2007 by Gurbanguly Berdymukhamedov, who has twice since been re-elected and has passed legislation to make himself 'President for Life'. Most importantly, his election marked a change in posture *vis-à-vis* Moscow, as the new president pursued an 'open door' policy and looked to develop new export markets for Turkmenistan's gas – to the east, in China, and to the south, in Iran. He was also courted by the EU who saw the vast resources of the region as potential feed gas for the development of a 'Southern Gas Corridor' that would reduce reliance on Russia.[35] Once he was in power, the 2008 financial crisis dealt a major blow to Turkmenistan's economy. The dramatic fall in oil prices resulted in Gazprom paying more for gas imports from Central Asia than it was receiving from exporting that gas to the EU. On 9 April 2009, an explosion on the CAC network stopped the flow of Central Asian gas to Russia.[36] Both sides blamed each other, but both gained from the cessation of gas exports. Gazprom no longer had to pay over the odds for gas that it did not need, and Turkmenistan was left free to pursue other options. That is not to say that the loss of revenue did not hurt the Turkmen economy, but China had already stepped in to provide financial support in the wake of the 2008 financial crisis, which, following the loss of gas export revenues, left Turkmenistan in a dire economic situation.

Against the backdrop of deteriorating relations with Russia, Turkmenistan, together with Kazakhstan and Uzbekistan, had negotiated an agreement with China to build the Trans-Asia-Gas Pipeline (now known as Line A). Prior to this, Turkmenistan had not been very forthcoming in terms of the conditions offered to foreign companies, limiting them to the technically challenging offshore fields in the Caspian Sea and refusing to get involved in any part of the value chain beyond the Turkmen border. The net result was that little progress had been made.[37] In 2007, China's CNPC (China National Petroleum Company) was granted a Production-Sharing Agreement to develop the onshore field at Bagtyýarlyk, which now supplies Line A. Financed by Chinese loans and built with the aid of Western and Chinese companies, the pipeline is 1,833 km long and was finished in record time. Line A started operation at the end of 2009 and was soon followed by Line B in 2010, bringing the total export capacity to 30 bcm. In 2013, Turkmenistan and China signed an agreement to increase exports to 65 bcm by 2020.[38] This required further upstream development to include the Galkynys field and the construction of Line C, with an additional 25 bcm, taking the total to 55 bcm. These deals allowed for 10 bcm of exports each from Kazakhstan and Uzbekistan. Thus, the actual export capacity available to Turkmenistan with the completion of Line C was 35 bcm. At the inauguration of the Galkynys field in 2013, it was announced that further development of the field would serve as the basis for Line D, with 25 bcm capacity, that would follow a different route through Uzbekistan, Tajikistan and Kyrgyzstan, acting as a force for much-needed regional integration. This was scheduled for completion in 2016–17 but work was suspended in 2017 and its current status remains unclear, though work may resume in 2020.[39]

At this point, it is important to stand back from the dizzying details of pipeline construction. As table 2.3 details, in the space of a decade Turkmenistan has totally reoriented itself away from Russia and developed new trading relations with China and, to a far lesser extent, Iran (more on the latter below). However, the country is still exporting significantly less gas than it did to Russia a decade earlier, in part because domestic consumption remains high. Have these developments

Table 2.3 Turkmenistan's gas exports, 2008–2018 (bcm)				
	Exports to Russia	Exports to Iran	Exports to China	Total exports	Total production
2008	42.30	6.50		48.80	61.6
2009	10.66	5.77		16.43	33.3
2010	9.68	6.50	3.55	19.73	40.1
2011	10.10	10.20	14.30	34.60	56.3
2012	9.90	9.00	21.30	40.20	59.0
2013	9.90	4.70	24.40	39.00	59.0
2014	9.00	6.50	25.50	41.00	63.5
2015	2.80	7.20	27.70	37.70	65.9
2016	1.10	6.70	29.40	37.20	63.2
2017	0.30	1.70	31.70	33.70	58.7
2018	0.00	1.90	33.30	35.20	61.5

Source: BP Statistical Review of World Energy, various years.

brought prosperity to Turkmenistan? First, the Chinese model has provided up-front loans to cover the cost of construction, but much of that money has probably disappeared into the pockets of the ruling elite, and that which has not has gone to pay foreign contractors – mainly Chinese – to develop the upstream capacity and build the pipeline infrastructure.[40] In a classic case of the 'resource curse', the billions invested have probably done little to improve the domestic economic situation. Furthermore, the revenue earned from gas exports to China now has to pay off the loans. Second, the oil-indexed nature of the agreements means that the revenue stream is subject to oil price volatility. Third, there are growing concerns as to whether Turkmenistan can meet its current obligations to China, let alone be able to deliver more gas. There are also problems with Uzbekistan's ability to deliver the required exportable surplus. The collapse in the oil price in 2014 was a major blow to Turkmenistan's finances, and a period of low prices will thus extend the time that it takes to repay its debts to China. In this context, it is perhaps not surprising that Line D ran into problems; furthermore, in the winter of 2017–18, Turkmenistan failed to meet its export obligations.[41] This has left China looking for alternative sources of supply, but where has it left Turkmenistan?

While geography is not destiny, for Turkmenistan it presents major challenges to getting its gas to market. As we have seen above, the country has stopped moving its gas north to Russia and has re-oriented its trade eastwards to China. But it also has plans to the south and the west. At the end of 2015, it completed the East–West Gas Pipeline to facilitate diversification.[42] Energy cooperation between Iran and the Soviet Union started in the 1960s, and at one point Iran exported gas to pay for military assistance. More recently, exports of gas from Central Asia to Iran have been used to compensate for the lack of domestic-capacity supply to the northern regions of the country. As table 2.3 shows, with the decline in exports to Russia, Iran is now Turkmenistan's second market, but – despite the opening of additional pipeline capacity in 2010 – the volume of exports has been modest, and, in recent years, there have been disagreements over payment. In early 2017, Turkmenistan argued that it was owed $1.5 billion for unpaid-for gas, which Iran disputed.[43] This trade is also part of a 'swap' arrangement that sees Turkmen gas supplied to Iran, which in turn exports gas to Azerbaijan. Iran receives a 15 per cent cash payment for this service. However, Azerbaijan has recently announced that, due to the development of domestic gas production, it stopped importing gas from Russia or Turkmenistan in 2018.[44] In a similar vein, Iran has now invested in the necessary pipeline infrastructure to supply domestic gas to its northern regions and has stopped imports.

Beyond modest trade with Iran, Turkmenistan also has grand visions to move gas south through the so-called Turkmenistan–Afghanistan–Pakistan–India (TAPI) Pipeline project.[45] Anceschi has described this as a 'virtual' pipeline that has little prospect of being built but is important to the current regime's vision of the future of Turkmenistan. At a more practical level, the security situation in Afghanistan is such that its role as a transit state is in question. Equally, both Pakistan and India have LNG as an option, and domestic exploration, as well as another equally dubious pipeline that involves Iran as a source of gas.[46] With the Russia option gone as a route to Europe, there is still talk of a Trans-Caspian Pipeline that would link to Azerbaijan and the Southern Gas Corridor – a route long favoured by the US and the EU. Until recently, the lack of agreement between the littoral states about the status of the Caspian Sea was a major stumbling block. But in August of 2018, agreement was reached to treat it as a sea, rather than a lake, and to grant common usage of the surface waters for all littoral states and to divide the seabed among those states. This enables the construction of a sub-sea pipeline, but Russia is likely to use environmental concerns to stall such a development for the foreseeable future. Equally, Azerbaijan wants to prioritize selling its gas to the EU, given the modest pipeline capacity that has finally been developed within the Southern Corridor. The Trans-Anatolian Natural Gas Pipeline (TANAP) that links Azerbaijan to Turkey will have an initial capacity of 16 bcm and the Trans-Adriatic Pipeline (TAP) that links it to Southeast Europe (including Italy) has a capacity of 10 bcm.[47]

The final twist in the tale is that, in late 2018, Gazprom announced that it would start buying Turkmenistan gas again, having stopped imports in 2016 stating that they had no commercial logic. The details remain unclear and the volumes are bound to be modest. Russia, having turned its back on Central Asia in 2009, may now be making a political gesture that recognizes the need to balance China's growing influence in the region through its One-Belt-One-Road initiative, but, as we shall see next, Russia and Central Asia are now also in direct competition for China's gas market.[48]

As the dust settles on what has been a turbulent decade for Turkmenistan, it has become clear that the Turkmen were never really able to dictate terms. The geopolitical economy of Central Asian gas means that Turkmenistan is dependent on its neighbours to provide secure and willing transit to distant export markets that, in turn, must be willing to provide the capital and security of demand needed to fund upstream development and the construction of the necessary transport infrastructure. China has clearly seized that opportunity, but on its own terms and for its profit, aided by the Turkmen kleptocracy. It is also happy to condone the lack of transparency that allows the political elites in Central Asia to profit from these deals. However, this has left Turkmenistan swapping a volatile, but at times profitable, relationship with Russia, for a more stable, but less lucrative, relationship with China that has locked them into a financial dependence that may well take a long time to turn a profit for the country, leaving them short of funds to develop new gas production and export capacity.[49] With the postponement of Line D and the loss of exports to Iran and Azerbaijan, Turkmenistan is left with a commitment to export 35 bcm to China, and little else, and it is currently struggling to do that. Whether or not Line D will be completed remains to be seen, but China may be well advised to look elsewhere for additional pipeline gas imports.

Russia's Eastern programme

Even in the Soviet period, there was always the intention to develop the resource wealth of the Far East to supply export markets in Northeast Asia. The regions to the east of Lake Baykal were considered too far from European markets to justify development for exports. Between 1974 and 1991, a railway – the Baykal–Amur Mainline – was built to link the region's resource hinterland to the Pacific coast for export. This included schemes to develop natural gas resources in Yakutia and oil and gas offshore of Sakhalin Island. The latter was the subject of an inter-governmental agreement with Japan in the late 1970s that saw exploration activity that identified significant resources.[50] This project was then the subject of US sanctions in 1979 to protest against the Soviet Union's invasion of Afghanistan. Japan's interest then dwindled as the investment environment deteriorated and oil and gas prices fell. However, the arrival of Mikhail Gorbachev in 1985 and the opening of investment opportunities sparked new interest in Sakhalin's offshore, this time from South Korean and US companies. In the late 1980s, a tender competition was held. Following the collapse of the Soviet Union, the new government of the Russian Federation awarded Production-Sharing Agreements to develop the region's offshore oil and gas potential to two consortia. Today Sakhalin-1 is being developed by a consortium led by Exxon Mobil, together with Russian state-controlled company Rosneft (and its local affiliate) and the Indian state-owned company ONGC (Oil and National Gas Corporation Limited) – this project has its origins in the original agreement with Japan. Sakhalin-2 is operated by the Sakhalin Energy Investment Company (SEIC) that is majority owned by Gazprom, together with Shell and Mitsui and Mitsubishi. The 1990s were turbulent times for both projects: they progressed at differing paces, with their ownership changing along the way. For the purposes of this discussion, the initial phases of development are of limited interest here as both pursued a strategy focused on finding oil rather than natural gas, but Sakhalin-1 did build a gas pipeline from the island to supply the Russian mainland. Later, following Gazprom's forced take-over of Sakhalin-2 in 2006, and at the decree of President Putin, Gazprom built a pipeline from Sakhalin to Vladivostok and developed a new offshore field – Kirinskoye gas condensate field – to supply the city.[51] The project was approved in July 2008 and opened on 8 September 2011, in time for the APEC (Asia-Pacific Economic Cooperation) summit at Vladivostok one year later. Built at great speed and great cost, the pipeline currently has a capacity of 8 bcm, with the possibility to be expanded to 30 bcm, but the full potential of the pipeline has been limited because US sanctions have impacted the ability of Gazprom to develop new offshore oil and gas production at the Yuzhno-Kirinskoye field.[52] However, there is the option to extend this pipeline to supply the Korean Peninsula and China. The construction of Russia's first LNG plant as part of the Sakhalin-2 project is of greater significance, but that is discussed in chapter 4. Meanwhile, interest in Sakhalin has waned as Gazprom has sought to develop its 'Eastern Programme' to monetize the gas resources of East Siberia and the Far East, and US sanctions are making the construction of a third train at their LNG plant difficult to deliver, adding to its troubled trading relations with the EU.[53]

Initial interest in gas exports 'east of Baykal' came from a joint venture between BP and the private Russian company Tyumen Oil Company (TNK) to develop the Kovytinskoye gas fields in the Irkutsk region, to supply a pipeline that would move gas south to Mongolia and China.[54] Like the Sakhalin projects, this project was then the victim of the rising tide of resource nationalism in Russia in the late 2000s, and its licence was revoked in dubious circumstances and then transferred to Gazprom. At the turn of the century, when Gazprom first talked about its 'Eastern Programme', it did not own any of the major gas fields in East Siberia and the Far East, but by the end of that decade – aided and abetted by the Russian state – it controlled most of them. What was lacking at the time was infrastructure, and a market to sell the natural gas, but over time those issues would be resolved. In May 2014, after years of protracted negotiations between Gazprom and its Chinese counterpart CNPC, a Sales and Purchase Agreement set up a thirty-year contract to deliver 38 bcm of natural gas a year from the Chayandinskoye and Kovytinskoye gas fields via the 3,000-km Power of Siberia Pipeline to the border at Blagoveshchensk (see figure 2.3). The Chinese had hoped to gain an ownership stake in those fields, as they had in Central Asia, but Gazprom was not prepared to surrender equity. The cost of the project to Gazprom is put at $55 billion, but the price of the gas was a sticking point for many years and

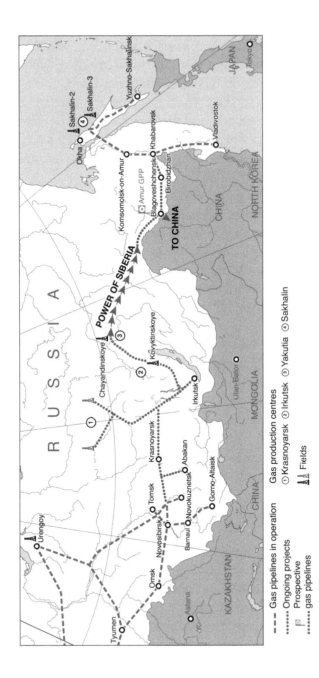

Figure 2.3 Gazprom's Power of Siberia Pipeline

Source: modified from Gazprom.com (2019), *Power of Siberia* [online]. Available at: www.gazprom.com/projects/power-of-siberia.

the final commercial terms are unknown. Gazprom wanted to link the price to the oil-index LNG price in Asia, but China considered this unreasonable and has struck a hard bargain. Gazprom's statement that the contract was for 1 tcm over the lifetime of the project and was worth $400 billion suggests a sales price of around 10–11 $/MMBtu, considerably lower than Asian LNG prices at that time.[55] However, as the price is oil-indexed, both sides are subject to future oil price volatility. Construction of the pipeline started in September 2014 and it has been in operation since December 2019, with volumes ramping up over the first few years.[56] For Russia, this project is of strategic significance as it promotes the development of East Siberia and the Russian Far East and also cements a new energy relationship with China, the world's most significant growth market for natural gas for years to come.[57] In addition, this development is important for Russia because it comes at a time when its relations with the United States and the EU are deeply troubled, with little chance of improvement on the horizon.

Russia actually proposed two projects to China: the Power of Siberia, and a second route known as the Altai Pipeline (sometimes called Power of Siberia 2) that would have connected to already developed fields in West Siberia – the same fields that supply the EU – with Western China.[58] Russia was clearly nervous about building a pipeline to a single market and the Power of Siberia project required the costly development of two new gas fields in very remote locations. However, the Chinese did not want to find their interests being used as leverage by Russia in its relations with Europe, nor did they want to pay Gazprom's European netback price for that gas, so they signed up to the Power of Siberia. Moreover, considering existing agreements with Central Asian producers, and considering that most demand growth is in the East and

Southeast of the country, China does not need additional deliveries of natural gas in the Western part of its country. Nonetheless, in November 2014, Gazprom and CNPC signed a Memorandum of Understanding on the construction of the Altai Pipeline. This pipeline would have a capacity of 30 bcm, bringing Russia's total pipeline export capacity to China to 68 bcm, significantly more than the 55 bcm provided by the three Central Asian pipelines. Until recently, progress on the Altai project has been limited, but a combination of problems with Central Asian supply and the postponement of Line D, surging natural gas demand in China, limited progress in expanding Chinese domestic production, and the trade dispute between the US and China, has resulted in renewed political support for the project in Beijing.[59] A third pipeline project bringing gas from Sakhalin Island has also been discussed, but this would require the development of new offshore fields. South Korea has long hoped that it could receive Russian pipeline gas via China, and natural gas is also seen as a non-nuclear solution to North Korea's energy needs.[60] As we shall see in chapter 4, China has rapidly expanded its LNG import capacity, but also sees benefit in expanding its strategic energy relationship with Russia. In less than a decade, China has become the centre of an expanding transcontinental pipeline system moving gas thousands of kilometres from Myanmar (8 bcm), Central Asia (55 bcm) and Eastern Russia (38 bcm), and, at a minimum, by the early 2020s China will be importing over 100 bcm of pipeline gas. The IEA predicts that China will soon become the world's largest gas-importer, with net imports reaching the levels of the EU by 2040. How China secures that gas is one of the most important issues facing the global gas market today.[61]

Conclusions

When moving large volumes of gas from producing fields to markets, large-diameter, long-distance pipelines represent an established cost-effective solution, and this is the way that most natural gas is moved within gas-consuming states. However, when it comes to exporting gas, the fixed nature of the pipeline infrastructure and the interdependencies that it creates can quickly complicate matters. As we have seen, geography, politics and economics can conspire in different ways to create a complex pipeline geopolitics that demonstrates different relationships between states and markets. In other words, traditional cross-border natural gas trade can come with risks, as we have seen in Central and Eastern Europe, and Central Asia. It is important to note that there are also examples of complicated geopolitical contexts where natural gas trade is not affected, e.g. Northwestern European trade with Russia, or shipments of natural gas from Qatar to the United Arab Emirates despite an ongoing diplomatic crisis.

The EU sees the solution to pipeline geopolitics in liberalization of domestic gas markets and their integration into a single European gas market. The belief is that a functional single market with a rigorous regulatory framework will protect member states against supply disruption. The problem is that the EU has limited power to impose its vision on member states – furthermore, only in some places is the physical infrastructure in place to underpin a single market. Behind the lack of progress is a difference in the ability and willingness of some member states to make the necessary reforms and investments. There remains a division between member states that have embraced liberal markets as a means to deliver energy security, and those that do not support that vision. Consequently, EU gas market integration is both a tremendous success story, and an ongoing tug-of-war between member states, EU institutions and other stakeholders at various levels of governance. The net result is that a single European gas market has yet to be achieved, and the complex geopolitics of gas is now serving to accelerate processes that may reduce further the EU's demand for natural gas in the future.

The situation is different at the other end of the Eurasian continent, in Northeast Asia. Over the last decade, a transcontinental pipeline system is emerging, focused on supplying growing amounts of gas to China. This has been a process dominated by state-to-state agreements that has been orchestrated by state-owned companies. The initial focus was Central Asia where China has cultivated a dependent relationship in its dealings with the Central Asian states, most significantly Turkmenistan. More recently, as the Chinese government has sanctioned a dash for gas to address the problem of urban air pollution, its attention has turned to securing additional pipeline supplies from Russia. Again, it has used Russia's relative political isolation to drive a hard bargain, whereby Gazprom is developing new, dedicated, gas fields in remote regions of Siberia and the Far East and building the Power of Siberia pipeline to deliver gas to the Chinese border. At the same time, as we shall see in subsequent chapters, China has expanded its LNG import capacity and is prioritizing development of its shale gas resources. Thus, while Gazprom would like to develop the Altai Pipeline, it is not clear whether China will need or want that gas. The focus of Chinese decision-makers continues to be diversity of supply. In addition – and unlike the EU – China benefits from central planning and decision-making. In the years ahead, assuming that Chinese policymakers will continue to view natural gas as an energy source that can help address problems of local air pollution, the country will need major investments in additional infrastructure to transport natural gas to market, and store it. The country will also need further market reform to allow for competition, both to grow domestic production and to facilitate diverse imports.[62] Contrary to in most of the EU, in China domestic exploration for natural gas ranks high on the agenda of state-owned enterprises, and policymakers. It is too early to conclude whether China will join the shale revolution, but as we shall see in the next chapter, it is trying. In addition, the country is diversifying its natural gas portfolio by investing heavily in LNG, which has emerged as a fuel for the future, as we discuss in chapter 4. We now turn to the first of two revolutions in global gas markets that have challenged more conventional natural gas trade and business practices, as described in this chapter.

CHAPTER THREE

The Shale Gas Revolution

This chapter considers the most significant development in the global gas industry in the last decade – namely the shale gas revolution that has dramatically changed the fortunes of the US from anticipating becoming a major LNG importer by the end of the first decade of this century to becoming a major LNG exporter a decade later. The combination of the established techniques of directional drilling and hydraulic fracturing, together with modern seismic surveying and reservoir modelling, has unleashed the wealth of natural gas resources that are trapped in the fabric of shale rocks. This chapter starts by examining the shale gas revolution in North America – it identifies the essential elements that explain its success, and reviews its wider impacts on the economy and the energy system of the US (acknowledging that shale gas production in Western Canada is part of the wider North American context). It then turns to the prospects for a wider global shale gas revolution by considering the current situation in Europe (with a focus on Poland and the UK) and China. The chapter concludes by considering the ways in which the shale gas revolution is challenging the established geopolitical economy of the natural gas industry.

The making of a revolution

The rise of (mostly) shale gas production in the United States has been revolutionary in that, until the late 2000s, it chiefly took place under the radar of conventional thinking and investment. Now, it is unquestionable that it has challenged conventional wisdoms within natural gas markets. As recently as 2006, US companies were constructing six LNG regasification terminals on and off US shores, expecting to import growing amounts of LNG to meet domestic demand for natural gas from 2008 onwards. Fast-forward a decade or so, and, according to the EIA, in February 2019 the US had been a net exporter of natural gas for more than twelve consecutive months.[1] A combination of increased seasonal pipeline exports to Eastern Canada, growing pipeline exports to Mexico, and slowly expanding LNG exports now more than outweigh falling pipeline imports from Canada and very modest LNG imports into the Eastern seaboard. According to BP's latest statistical review, net exports – the volume of exports over imports – stood at 16 bcm in 2018.[2] We discuss these and other long-term impacts of the rise in shale gas production in this chapter, but first we ask the question why these events took place in the United States, and not elsewhere. In turn, we explore whether other 'revolutions' are around the corner – and, if so, where. We find that, rather than being an overnight sensation, this revolution has been three decades in the making, and that a complex and likely unique set of factors has allowed US companies to flourish, and turn the domestic energy landscape upside down, with ripple effects around the globe.

It has been known for many decades that large amounts of methane are trapped in rock layers all around the world, but the challenge has been to find ways to extract it at reasonable cost.[3] In the United States, over time, it became clear that the geology in several parts of the country was very favourable to resource extraction – with current technology. The latter is important, because, prior to 2003, the Marcellus shale in the Northeast of the country was considered an organic-rich rock, but one that drew limited commercial attention, because the permeability of the rock was typically low. When companies started to combine hydraulic fracturing and horizontal drilling (both existing industry practices), this changed, and following on from early development in the Barnett shale (in Texas) and the Haynesville shale (in Northwest Louisiana, Southwest Arkansas and Eastern Texas), they soon moved to the Marcellus (in Ohio and Pennsylvania, but not New York State), which has subsequently developed into one of the largest natural gas fields in the world. Underlying the Marcellus is the Utica shale, which is at greater depth and therefore received less attention at first (drilling deeper essentially means higher extraction costs), but this changed from 2010 onwards as companies learned more about the rock. It turned out that, besides methane, significant portions of both shales contained large amounts of marketable gas liquids like ethane and propane, which often became the chief incentive for companies to keep drilling. Geologists can learn a fair amount about the subsurface using 2D and 3D seismic surveys, but collecting data by drilling is the only way to understand clearly the geological conditions in a given location. In the US, thousands of wells are drilled annually, which has allowed companies to estimate better the size of the resource base, but also to be able to aim for so-called 'sweet spots' when prices are low, and profit margins are squeezed. Though not the topic of this book, as we noted in the introduction, a large part of the US onshore industry is now focused on producing oil from shale rock layers (generally known as 'tight oil', though at times erroneously referred to as shale oil),[4] which also has major consequences for natural gas production in the country, because most of that tight oil comes to the surface with significant amounts of methane, or 'associated gas', which is essentially a by-product at very low – and at times even negative – cost (the problem of flaring was highlighted in chapter 1). In sum, many decades of drilling in the US (and parts of Canada) have revealed geological conditions extremely favourable for hydrocarbon extraction, with a variety of consequences that are discussed later in this chapter.

An obvious, but important, prerequisite for a burgeoning industry is the availability of infrastructure to bring natural gas to market. The United States has historically been one of the larger producers of (conventional) natural gas in the world, and as a result has a very well-developed physical infrastructure to support natural gas transportation and trade. Today, there are over 300,000 miles of interstate and intrastate transmission capacity in place, with over 1,400 compressor stations, 400 underground gas storage facilities, and several dozen import and export locations – by both pipeline and liquefaction – and regasification facilities.[5] It is, by all standards, the world's most developed natural gas market. This assists production companies and traders, who often do not have to worry about getting their product to market. To be sure, the advent of shale gas in areas of the country where in recent years little or no extraction had taken place (e.g. the Marcellus and Utica) has required that new infrastructure be built to serve clients. This build-out can be controversial (because people do not want infrastructure development close to their home, or fear being locked into long-term high-carbon diets), but the data suggest that most proposed projects are eventually built. Between 2009 and 2018, 239 new pipelines, expansions and reversals were proposed and completed nationwide.[6] Regarding access to this infrastructure, a critically important factor that enabled the surge in natural gas production is that capacity rights (the right to ship natural gas through a given pipeline) are by law unbundled from infrastructure ownership. This is important because it prohibits owners of infrastructure from withholding capacity, thus restraining competition and market growth. Some scholars have called this the most underappreciated factor that helped enable the rise of shale gas production in the United States.[7]

For non-US observers, it is important to keep in mind that most natural gas production is confined to a limited number of states (i.e. Texas, Oklahoma, Louisiana, New Mexico, Colorado, Pennsylvania and Ohio).[8] The rapid rise of the Marcellus shale basin is demonstrated by the fact that, back in 2008, Pennsylvania and Ohio accounted for less than 1.5 per cent of US natural gas production, but by 2017 that share had increased to 26 per cent, and during the same period total US production increased by 35 per cent. Most of these regions have a long history of onshore hydrocarbon extraction, and therefore the industry mostly enjoys public support for its work: for better or worse, people have grown accustomed to resource extraction, and often benefit directly or indirectly from it. In states in the Northeast of the country, where in recent decades natural gas production had been in decline, the influx of activity and capital has generally been welcomed by state legislatures, though there are important exceptions, and it is a mistake to think that the natural gas industry has enjoyed support throughout the United States.[9] We would point to continued controversy and local opposition to the build-out of natural gas infrastructure into the Boston area, where, as a result, peak winter demand is met with imports of LNG and switching to fuel oil. In the winter of 2018–19, the ultimate irony was that the first production from the Yamal–Europe project on the Arctic coast of West

Siberia in Russia ended up in Boston![10] We could also point to the 2014 ban on hydraulic fracturing in New York State, despite the likelihood of significant natural gas finds in its subsurface.[11] As mentioned earlier, there is also controversy about the flaring of major amounts of natural gas in North Dakota in 2013, when oil production boomed and there was not enough infrastructure in place to bring the associated natural gas to market, which thus had to be vented, or flared. In October 2015, methane (and some ethane) started leaking out of Aliso Canyon storage facility in California, and it took SoCalGas until February of the next year to seal off the leak, making this, with an estimated 100,000 tons of methane, the largest natural gas leak in US history.[12] One of the consequences of this incident has been a reorientation of state legislatures to support the deployment of alternative renewable energy technologies, in a move to reduce the share of natural gas in the state in the long term.[13] Consequently, there are a significant number of states that are critical of fossil fuel extraction and want a swifter transition towards a low-carbon economy, and away from natural gas. What has undoubtedly helped garner support at the state level is the unique mineral rights system in the United States. It is the only country in the world where the owner of the land also owns the resources underground (though the two can be separated). Essentially, it means that a developer can knock on any given door and negotiate permission to drill on someone's land. As royalty rates and signing bonuses went up, access to land often became easier, as local communities benefitted financially from the influx of the industry. However, this also became a source of inequity and division in some communities and the stuff of Hollywood movies (the 2012 film *Promised Land*). Similarly, in states where water can be scarce, such as Texas, access to water rights has been negotiated with farmers and land-owners, allowing the industry to continue operations, at the right price.

Contrary to most other parts of the world, North America, and the US specifically, has a very competitive onshore oil and gas service industry. At the end of August 2019, according to Baker Hughes, 878 drilling rigs were operational on land in the United States (compared to 1,028 a year earlier), 784 rigs were drilling horizontal wells, but only 162 were dedicated to gas production.[14] This compares to a total of 1,162 drilling rigs that were active in the rest of the world. Not only is the equipment available, but there are also hundreds of small- and medium-sized petroleum companies vying to use their equipment to extract oil and natural gas. This is important because these wildcatters were the catalysts of the fracking boom, rather than the larger companies and the international oil companies (IOCs). The latter were relatively late to appreciate the magnitude of this phenomenon, as illustrated by decisions to invest in LNG regasification terminals in 2006 (to import natural gas). The takeover of XTO Energy by Exxon Mobil in December 2009, for $31 billion, plus $10 billion in XTO Energy debt, confirmed its shift towards more domestic production and is a significant milestone in the evolution of the industry. Fast-forward, and the 2019 takeover of Anadarko Petroleum, with its major acreage in the Permian Basin in West Texas, by Occidental Petroleum illustrates the long-term commitment to producing domestic oil (and lots of associated gas) that these companies are planning to make.

A romantic narrative would have one believe that the shale revolution was a classic success story of US private-sector ingenuity. Undeniably, entrepreneurs such as George P. Mitchell played a major role in the breakthroughs in the 1980s and 1990s that led to the surge in domestic natural gas production in the following decades. Yet it is important to acknowledge the substantial political support and favourable fiscal and regulatory climate for the industry as well. In the 1970s, fundamental research was supported financially through the Gas Research Initiative and the Department of Energy's National Energy Technology Laboratory.[15] It had long been known that large amounts of methane were trapped in shale rock layers, and the turbulence of the Oil Crises provided the government with a good reason to fund companies to experiment with technologies to extract those resources, even though at the time few, if any, believed that this would be commercially viable. When modest progress was made, resources produced from (Devonian) shale received significant benefits through wellhead price deregulation, and tax credits under the Natural Gas Policy Act. In 1980, a Section 29 production tax credit for unconventional gas was created, which lasted until 2002. Consequently, when the price for 1,000 cubic feet of natural gas hovered between $1.50 and $2.50, under the Crude Oil Windfall Profits Tax Act, companies enjoyed an incentive of $0.50 per 1,000 cubic feet of natural gas produced from shale rock. Relatively strong political support can also be found at the state level. In states such as Texas, regulatory authorities have historically been closely aligned with the industry. It is fair to say that, in the early days of the fracking boom, regulatory authorities and legislatures were also playing catch-up with the pace of development. Yet regulatory regimes matured, even though major differences between key states remain.[16]

The existence of a large gas market was not just important for the availability of infrastructure. It also allowed for very significant growth of demand for natural gas (and associated liquids), first predominantly in electricity generation, but also as a feedstock in industry, and for transportation. Access to cheap capital has also been a critical driver of continued investment in new drilling activity. The 2008 financial crisis was followed by historically low interest rates, allowing companies to keep borrowing money to finance new operations, whereas under normal circumstances they would have financed those out of their own cashflow. It is for this reason that some analysts have expressed scepticism about the longevity of some of the drilling activity in the United States, even though that ignores the wide disparity between various companies in the country.[17] When prices for both natural gas and oil crashed in 2013 and 2014, it forced companies to reassess their spending and improve the efficiency of their operations. Undoubtedly, many came out stronger, even though reports of money flowing back from companies to financiers are still the exception rather than the rule, and the industry remains cash-negative. The British energy economist Paul Stevens, when reviewing the US shale revolution, identified no fewer than seventeen characteristics that explain the US shale phenomenon.[18] Most of these have been discussed above and there is no need to provide a detailed enumeration; the point is that a complex set of factors came together to create the conditions for the rapid development of the shale gas industry in North America, which may make it difficult to replicate elsewhere.

The direct and indirect consequences of the shale revolution

Assessing the direct and indirect consequences of the rise in natural gas production in the United States resembles a Chinese firework display: once the smoke has cleared, spectators assess the new moment, and whether everything is still in place. In this instance, that is unequivocally not the case: the shale revolution has fundamentally changed the energy landscape in North America and beyond. However, some of the consequences have been more difficult to quantify than others, while others have been grossly exaggerated. In this section, we follow the historical timeline as we offer our assessment of the economic, environmental and geopolitical consequences. We end with a characterization of the uncertainties that lie ahead, before we assess the potential for unconventional natural gas to go global.

The surge in natural gas production starting in the late 2000s could arguably not have come at a better time. As the United States was recovering from the greatest recession in decades, the booming oil and gas industry was arguably a positive economic story amid downturn and significant unemployment. Some studies estimated that continued growth in oil and gas production could contribute 0.2 per cent per year to GDP between 2013 and 2020.[19] By now, that growth has slowed, also because other parts of the economy have recovered, employment numbers are back to (very) low levels, and capital can be invested elsewhere, too. It is worth noting that the economic impact at the state level can be more significant than macro numbers may at first sight suggest. States such as North Dakota have by now gone through typical boom-and-bust cycles, with peaks (in 2014, a typical 700-square-foot one-bedroom apartment cost $2,394 per month, over $800 more than a similar unit in New York, Los Angeles or Boston) and lows (in 2015, North Dakota was the only US state to report a significant increase in its unemployment rate, as oil prices had started to tumble in June 2014).[20]

The growth in domestic natural gas production had indirect impacts outside of North America. Whereas the market had been anticipating a rise in imports of LNG in the second half of the 2000s, demand was fulfilled with domestic production instead. It resulted in excess supply in the market in 2008/9, and regasification projects in the United States were delivered but then stood idle. Initially, cargos from various parts of the world – e.g. Nigeria, Qatar – destined for the United States increasingly found their way to Asian markets, but when demand could not keep pace, the European market became the 'sink' for LNG supply.[21] This had a downward effect on prices, and incentivized incumbent companies to reconsider their gas pricing strategies in (parts of) the continent. Granted, sometimes with a regulatory push, companies like Dutch GasTerra, and Statoil from Norway moved away from the traditional indexation of gas prices to oil, and instead based their contracts on gas-to-gas prices of NBP in the UK and TTF in the Netherlands. This sparked a trend that continues to date. Not all companies have moved at the same pace – Algeria's Sonatrach and Russia's Gazprom still employ hybrid pricing systems, with a significant role for oil prices, but they too, reluctantly, are moving away from their traditional business practices, and it is only a matter of time before all prices in Europe are based on supply-and-demand dynamics for natural gas, not other commodities.

Electricity generation is the chief source of natural gas demand growth in the United States. It is one of the few markets where fuel sources compete head to head, essentially based on the marginal cost (though somewhat constrained by the available generation capacity). When coal prices are low, it tends to trump natural gas in the merit order, and vice versa. As domestic natural gas production grew, prices fell, and the fuel became more attractive for utilities. Between 1949 and 2005, coal was used to generate around half of the electricity used in the United States. By 2017, its market share had fallen to an estimated 30 per cent, and, for the first time in history, natural gas was the main feedstock for electricity generation, at 34 per cent (it held about 20 per cent market share in 2005).[22] It is important to note that other factors played an important role in coal's demise, most prominently the growth in investments in solar photovoltaic (PV) and onshore wind capacity, and regulatory constraints such as the 2011 mercury rules (which the Trump Administration is trying to roll back). Yet the growth of natural gas continues, with the EIA suggesting that, between 2018 and 2035, another 89 gigawatts (GW) of coal-fired generation capacity will be retired, to be replaced mostly by natural gas.[23] It is also the case that no new coal-fired power plants were being built in the US in 2019.[24] To be sure, these are not one-on-one substitutions, and old natural gas-fired power plants are retiring as well, but the overall trend is clear: natural gas and renewables are increasingly dominating the electricity-generation mix in the United States. The consequences of this shift have been well documented: it was the key contributor to the sharp and unprecedented drop in US carbon emissions, starting in 2007.[25] It is important to note that other factors played almost equally important roles in this transition, most prominently the growth in renewables, and a decline in demand, for both structural reasons (efficiency) and temporal economic ones (recession). Going forward, it is anticipated that growth of natural gas in power generation will continue, though at more modest levels.

The rise of natural gas has not been without controversy. The movie *Gasland* suggested potential negative environmental impacts of shale gas development. As drilling intensified, incidents did happen, and these received broad media attention and regulatory scrutiny. Concerns about hydraulic fracturing and directional drilling mostly focused on fugitive methane, water pollution and induced seismicity (we discuss these concerns in more depth below). Fugitive methane is gas that is leaked at any point during the production cycle, from production to combustion or processing. It is a relatively poorly understood phenomenon that is the source of considerable disagreement, chiefly because it is difficult to measure and quantify. A study by the US Environmental Protection Agency (EPA) of trends in methane emissions from the US gas system between 1990 and 2017 found significant emissions reductions from the midstream and downstream (e.g., due to use of better materials), but these reductions had in part been offset by an increase in upstream emissions linked predominantly to oil and gas production from shale.[26] Since 2018, further studies have emerged suggesting that emission rates may in fact be much higher and that shale gas development in the US is making a major contribution to a surge in global methane emissions, calling into question the role that natural gas may play in a GHG-constrained world.[27] We discuss this topic in more detail in chapter 5, but suffice to say that our understanding of methane emissions in natural gas production urgently requires further improvement, but it is still the case that natural gas is a cleaner fuel with lower carbon emissions than coal when used to generate electricity. In addition, there is broad recognition of the fact that, with existing technologies and against modest cost, a major share of methane emissions can (and should) be further curtailed. Industry opinion on whether regulation is needed to enforce usage of those technologies is not uniform. When, in August 2019, the Trump Administration tried to rollback Federal regulations on methane emissions, some in the oil and gas industry, such as Exxon Mobil, Shell and BP, criticized the move, fearing that it would damage the environmental case in favour of natural gas.[28] In the coming years, further research will be carried out that will improve our understanding of the problem. In addition, forward-leaning companies will implement best practices on their own initiative. Other companies will be less willing to do so, but will be forced to by regulation, shareholder pressure and/or financial institutions. In some parts of the world, such as the EU, more stringent regulation will likely be proposed soon. It is also worth noting that it is the controversy surrounding fugitive methane emissions in the shale industry that has raised the profile of the role of methane more generally as a GHG, and this is resulting in a more holistic approach to monitoring and mitigating the emission of this significant cause of climate change.[29]

Pollution of ground water with fracking fluids has been reported in a couple of individual cases, though this continues to be disputed.[30] With the available evidence, it is safe to conclude that in the overwhelming number of cases well drilling and completion is done properly, without risks to ground – or drinking – water. In individual cases, mistakes can be made, and there is no way of ruling that out with 100 per cent certainty in the future, other than to stop producing natural gas. In North America, induced seismicity has mostly, but not exclusively, been linked to the reinjection of flowback water into wells, but in isolated cases – such as in northern British Columbia in Western Canada, or Oklahoma – also to hydraulic fracturing itself. Parts of Oklahoma have witnessed a major increase in seismic activity, and the severity thereof, though mounting concerns have had a modest impact on drilling activity in the state.[31] Building new gas pipelines in the United States is increasingly controversial, illustrating both classic NIMBY behaviour, and also a (likely growing) dissatisfaction with the pace of the transition to a low-carbon economy. In short, there are genuine concerns about the cumulative environmental impact of shale gas development in the United States. A part of the initial concern must be linked to the speed with which the industry evolved, and the catch-up required of regulatory authorities, legislatures and science. Over time, and in aggregate, the local environmental risks stemming from natural gas production seem manageable, and the benefits to the wider economy and society are significant. To be sure, that does not mean that there will not continue to be environmental concerns linked to natural gas production, and consumption, going forward, but smarter regulation and other incentives can help nudge all producers to continue to improve best practices. As well as shorter-term risks, there are longer-term uncertainties worth considering. Specifically, the question is open as to what role natural gas may play long term, in scenarios of deep decarbonization.[32] We return to this question in chapter 5.

With production growth, the certainty about the longevity of the shale phenomenon grew as well. Initially, production outlooks were understandably modest, as geologists were hesitant to extrapolate recovery rates to other natural gas 'plays'. In addition, it had become clear that, after an initial peak in production, once gas was flowing up a well, a significant drop in production quickly followed. This in turn was an incentive for the industry to continue drilling new wells, as were licence requirements to drill a given number of wells – contrary to conventional gas production, where a couple of dozen wells could sustain production for years, if not longer, depending on the size of the resource. The revolution that was unfolding did not escape the attention of the federal government under the Obama Administration, either. For instance, it attempted to regulate some of the environmental concerns linked to hydraulic fracturing, with limited success. At the same time, it benefitted from the positive consequences of extensive fuel-switching on the nation's carbon footprint. In his first term, President Obama endorsed the oil and gas industry, and its contributions to US energy production and its improved GHG footprint, unequivocally.[33] We have only anecdotal evidence to support this theory, but feel comfortable saying that, without the contributions of natural gas, the US government would likely not have been so forward-leaning in the years prior to the signing of the Paris Climate Accord, in December 2015. The United States had an easy argument to make: 'use relatively cleaner fuels to displace more polluting ones, like is happening in our country, and without major government intervention'. It is worth adding that President Obama's views about natural gas became more nuanced over time, likely for a variety of reasons. First, there are significant uncertainties about the role that natural gas can play in the long term (without further curtailing methane emissions, and commercializing CCUS technologies in electricity and industry); and, second, a part of the electorate views natural gas as just another fossil fuel, and is more in favour of keeping the resource in the ground, rather than producing it.[34]

Natural gas producers in the US were soon in search of new markets in which to sell their product. Most prominently, Mexico emerged as a major new market, because of a significant decline in its domestic production of (associated) natural gas. Following the opening of the Mexican market for foreign direct investment in 2013, several pipeline projects were proposed, financed and completed, and within five years, exports of US natural gas into Mexico rose to reach 2.2 tcf in 2018.[35] It is worth noting that, before the pipelines were completed, Mexico also figured as the number one destination of LNG from the US.[36] As more pipeline capacity comes online, less LNG imports are expected, even though downstream infrastructure constraints may continue to prevent natural gas from flowing to market here and there. There are some uncertainties regarding Texan–Mexican natural gas trade going forward, most prominently the desire of the Mexican government to boost domestic production at the expense of imports, though the current government is not allowing domestic shale gas development. However, for the time being, natural gas from the US is flowing into Mexico in very significant quantities, and these numbers are expected to grow as more infrastructure is built, both on the US and Mexican sides of the border, eroding the need (and incentive) to invest in Mexico's domestic resource potential. To the north, where Canada has been a major producer of natural gas itself, the consequences of US production expansion have mostly been felt by Canadian producers, who have seen their main export market evaporate. Canadian exports into the US are unlikely to disappear entirely, mostly because infrastructural constraints will not allow US natural gas to flow to all corners of the country, but Canadian producers will need to find alternative markets to sell their produce. In British Columbia, the hope is that their shale gas resources can feed the development of a new LNG export industry on the Pacific Coast that will compete with new US exports in Asian markets. In addition to regional exports by pipeline, exports of natural gas in the form of LNG have become ever more attractive.

Conventional wisdom had dictated that the US would become a major importer of LNG, but when spot prices fell from $12.69 in the summer of 2008 to $2.99 by September of 2009, Cheniere Energy was the first to propose that their newly built regasification terminal should be converted into a liquefaction plant, so that US natural gas could compete on international markets.[37] More details, and the long-term and fundamental consequences of these developments for the LNG market, are discussed in the next chapter.

Because of the boom in natural gas (and tight oil) production, a renaissance in manufacturing in the United States has slowly but surely materialized. It is important to note that relatively small amounts of natural gas are used in the industrial sector (an estimated 3 per cent of total production in the US), though the fuel is used as feedstock for nitrogenous fertilizers, and chemical products including ammonia, hydrogen and methanol. However, NGLs like ethane have historically formed the basis of elementary petrochemical and plastics production in

North America, unlike in East Asia and Europe, where chemical producers primarily use naphtha (from oil refining) as a feedstock. Since many newly drilled wells in the United States contain wet gas (including large amounts of liquids), availability and pricing of NGLs has been favourable, and, as the certainty about the longevity of the shale phenomenon grew, so did investor confidence regarding new manufacturing facilities. Consequently, since 2010, 334 new chemical-industry manufacturing projects have been announced, valued at over $204 billion, of which 53 per cent are either completed or under construction.[38] It is also worth noting that over two-thirds of these investments are considered foreign direct investment, or include a foreign partner.

A final and less tangible consequence of the shale revolution has been a political one – namely, the changing posture of the US diplomatic corps. From 2010 onwards, the State Department supported a series of initiatives that essentially served the purpose of providing technical assistance (helping officials in host countries better understand their resource potential) and sharing best practices (inviting delegations from foreign countries to come and learn about US state experiences regarding shale gas development). Starting as the Global Shale Gas Initiative, these efforts were later renamed the Unconventional Gas Technical Engagement Program, and eventually, together with other programmes, morphed into the Bureau of Energy Resources (ENR) within the State Department, which by 2015 had a budget of over $16 million, and more than ninety employees in 2016. In a major address that the then US Secretary of State Hillary Clinton gave at Georgetown University in 2012, she laid out a broad vision of what 'energy diplomacy', in her view, could mean for the United States.[39] The newly found natural gas and oil riches featured very prominently and were accredited with many possible advantages for the United States, including alleviating energy poverty, helping with conflict resolution, freeing allied countries of the yoke of dominant gas suppliers, enhancing free trade, and promoting the decarbonization of the energy system. As described earlier in this chapter, some of these consequences of the expansion of natural gas production are very valid observations to make, yet the role of the federal government herein is much more complicated to pin down. By now, some work has emerged that nuances the role that diplomacy plays in US energy markets, which are predominantly driven by private-sector investments.[40]

It is hard to overstate the significance of the shale gas revolution in the US, especially as those same technologies have also been used to bring about a similar renaissance in onshore oil production, which in combination has totally transformed the position of the US in the global energy system. Andreas Goldthau has suggested that the US shale revolution has brought a 'triple premium': 'low energy prices and an economic boost; a reduced carbon footprint underpinned by simultaneous economic growth; and a gain in international security and geo-economic standing'.[41] However, it is not all good news: oil and gas production is also having a significant environmental and social impact on the communities where it is most prevalent. Thus, together with macro-economic and geopolitical benefits, there is also the potential of significant costs at the local scale. One key difference in the US is that private ownership of subsoil rights means that individuals in the community benefit financially, but, as we noted above, that too can become a source of conflict at the local level. The final factor to consider is that the nature of shale production means that the industry needs to keep drilling to maintain production, and to date mostly relies on debt finance to fund drilling. Consequently, and despite the rapid evolution that the industry has gone through in the last decade, there are ongoing questions about the environmental and financial sustainability of this industry. Nonetheless, there can be no denying that the US now benefits from a plentiful supply of natural gas, which, even at 5 $/MMBtu – it is currently well below $3 – is cheap by international standards.

In sum, we can say that the essential success factors behind the shale revolution come down to a favourable geology, initial government support, available infrastructure and a market, amenable governance and regulation, access to capital, private ownership of mineral rights, and significant social acceptance. Below, these success factors are used to access the prospects for shale gas development elsewhere in the world.

Will the revolution travel?

Our assessment of the prospects for shale gas development outside North America starts with geology. The first thing that should be said is that, in almost all instances, estimates are just that – estimates of technically recoverable shale resources. These numbers are highly speculative, and it is only after extensive exploratory drilling and appraisal that more accurate resource estimates can be made, and, even then, further drilling and flow testing is necessary to determine whether or not there is an economically viable proven reserve. In this context, it is worth remembering that it took thirty years or more for the US shale revolution to become a reality. As we shall see, there is only so much that can be learned from the US experience, and expectations of simply transplanting the US model are therefore unrealistic.

In 2011, the EIA released its first World Shale Resource Estimates, with updates in 2013 and 2014.[42] The estimates cover forty-one countries. Table 3.1 presents the top ten

Table 3.1	Top ten countries with technically recoverable shale gas resources (EIA 2013 estimates)			
Rank	Country	Technically recoverable shale gas reserves (tcf)	Technically recoverable shale gas reserves (tcm)	Percentage of world total (%)
1	China	1,115	31.6	15.3
2	Argentina	802	22.7	11.0
3	Algeria	707	22.0	9.7
4	USA	665	18.8	9.1
5	Canada	573	16.2	7.9
6	Mexico	545	15.4	7.5
7	Australia	437	12.4	6.0
8	South Africa	390	11.0	5.3
9	Russia	285	8.1	3.9
10	Brazil	245	6.9	3.4
	Europe	471	13.2	6.4
	Poland	146	4.1	2.0
	France	137	3.9	1.9
	World total	7,299	206.6	98.4

Technically recoverable: the volumes of natural gas that could be produced with current technology, regardless of prices and production costs.

Source: available at www.eia.gov/analysis/studies/worldshalegas.

estimated shale gas resource holders, plus details about Europe, and specifically Poland and France (the two EU member states with the most significant resource estimates).

The first thing to note from the 'Top Ten' is that most of them already have a significant oil and gas industry. This is important, because exploring their shale potential may consequently not be an industry and/or policy priority. Russia is a good example. The following discussion focuses on the three countries where there has been significant activity in relation to shale gas development: Poland and the United Kingdom (as part of a broader discussion on the EU), and China. To be sure, there is also considerable activity elsewhere – for instance in Argentina, which exported its first cargo of LNG in July 2019.[43] However, we focus our analysis on China and the EU because they are the two largest import markets for natural gas, and domestic shale gas production would change their gas balance.

Shale gas in Europe

Given what was said in the previous chapter about the geopolitics of pipeline gas in Europe, it should come as no surprise that the EIA's resource estimate prompted considerable excitement in parts of Europe. As a major gas-consuming region with falling domestic conventional gas production and increasing import dependence, the prospect of a new indigenous source of natural gas offered a solution to Europe's energy security concerns.[44] However, support, from both government and society, was not universal and from an early stage there were member states that decided to ban hydraulic fracturing, notably France and Bulgaria, and later Ireland. Elsewhere – for instance, in Germany and the Netherlands – governments decided to commission studies to understand better the environmental concerns linked to shale gas extraction, often then to ban hydraulic fracturing after several years.

The European Commission was rather muted in its support for the industry. Following the IEA's publication of its *Golden Rules* that suggested a checklist for responsible development of shale gas, the Commission considered an EU Shale Gas Regulation, a move that was resisted by Poland and the UK; instead, it came up with a set of best-practice recommendations.[45] It subsequently tested member state compliance against those recommendations and funded further research, and tasked the EU's geological surveys with coming up with fresh resource estimates, which it published in 2017.[46] The negative image of shale gas development, inspired in large part by the *Gasland* movie, together with highly effective environmental movements in many member states, meant that shale gas development was quickly associated with negative local environmental impacts, and its development was seen as incompatible with the EU's climate-change ambitions. It has been suggested that the precautionary principle, enshrined in the Lisbon Treaty, helped set the tone for hydraulic fracturing in the EU, as it would with any new technology.[47] More generally, the campaign against shale gas across Europe tarnished the natural gas industry's image as a cleaner alternative to coal. There were only two member states where shale gas development gained momentum: Poland and the UK.

Poland's Shale Quest
The positive EIA resource estimate for Poland generated considerable optimism, even though in 2015 its own geological survey produced a lower set of estimates of total shale gas deposits of 1,920 bcm and recoverable gas at 346–768 bcm.[48] This is still a significant amount when one considers that in 2018 Poland's total gas consumption was a modest 19.7 bcm, and domestic production was 4.0 bcm; thus, 15.7 bcm was imported, largely by pipeline from Russia.[49] Thus, domestic shale gas production could significantly improve Poland's gas security. Furthermore, Poland is highly dependent on domestic coal production, which in 2018 supplied 47.5 per cent of its primary energy and 79.2 per cent of its electricity. This puts it at odds with EU's climate ambitions, and access to domestic gas could certainly aid in the decarbonization of its energy system.

In this section, we analyse Poland's shale gas experience. First, there is the matter of the geology: both the EIA estimate

and subsequent work by the Polish Geological Institute identified significant resources in place. The challenge then became
drilling the 200–300 wells needed to assess the potential for
commercial development. For reasons discussed below, this
never happened, and by 2015 when exploration activities
effectively stopped, only 70 wells had been drilled. Poland's
state-owned companies – PGNiG and PKN Orlen – led the
way, in cooperation with North American companies that
included ExxonMobil, Marathon Oil, Talisman, Halliburton,
Chevron and ConocoPhillips, as well as majors such as Total
(France) and ENI (Italy). In retrospect, we learned several
things. First, due to the higher clay content in the Polish shales,
the techniques that had been so effective in the United States
were not so in Poland. Second, the strong support of the Polish
government turned out to be problematic, in that it counted on
a shale boom and placed too heavy a tax burden on the early
exploration activity, making investors nervous. Third, given
the dominance of coal in the energy mix, Poland at the time
had a relatively under-developed natural gas market, which
needed significant market reform, and investment in infrastructure.[50] The fourth factor of governance and regulation
proved a major hindrance to the industry. The Polish system
still bears the hallmarks of its socialist past, with a complex
and inflexible state bureaucracy with overlapping responsi-
bilities between national ministries that failed to cooperate. It
proved difficult to design an attractive fiscal regime to attract
investment, in part because policymakers seemed overly
focused on taxation of what they believed was soon going to
be a burgeoning industry. Add to this the cost of bureaucracy
and the disappointing drilling results and it is no surprise that
many of the foreign companies exited within a few years of
arrival. The crash in the oil price in 2014 was the final nail
in the coffin as it provided a good reason to withdraw and

focus on more profitable opportunities elsewhere. Ironically, and unlike in most other EU member states, public acceptance for shale gas was high in Poland's case. This was largely based on energy security concerns and the perceived economic benefits. In the end, the Polish shale failure was a combination of below-ground technical problems and above-ground governance and regulation issues.[51] Following this disappointment, the Polish government seems determined to improve its gas security through other strategies: building an LNG import terminal, investing in the Baltic pipeline project via Denmark to access Norwegian gas, and improving its interconnection to the Central European gas network. Its current long-term contract with Russia's Gazprom expires in 2022 and it is determined not to renew it. The problem with this strategy is that Poland will likely be stuck on a high-carbon diet for many years to come as it cannot move away from coal: security concerns trump all other policy objectives in Warsaw.

The UK 'Going All Out for Shale'[52]

The UK has been a significant oil and gas producer for decades, but although over 2,000 oil and gas wells have been drilled onshore, almost all its production has come from offshore fields in the North Sea. The EIA's reserve estimates put technically recoverable shale gas resources at a very modest 740 bcm. The UK's British Geological Survey (BGS) has conducted 'gas-in-place resource assessments' for four regions, and the Bowland-Hodder shale formation that straddles the North of England has the best prospects. The BGS central estimate for the Bowland-Hodder was 37.6 tcm and the range was between 23.3 and 64.6 tcm. This is a significant resource when one considers that in 2018 the UK consumed 78.9 bcm of gas. Production from the North Sea peaked in 2000, and by 2004 the UK had become a net importer of natural gas. Today, it imports about half of the gas that it consumes, with the majority coming directly by pipeline from Norway and the remainder from continental Europe via two interconnectors and by tanker to three LNG import terminals. Current estimates from the UK's Oil and Gas Authority suggest that offshore production will fall more rapidly than demand, and that by 2035 the UK could be importing 72 per cent of the gas it consumes. In this context, the prospect of a significant amount of gas in place has stimulated both government and industry interest.[53] The UK has no national oil company (NOC) – instead, its oil and gas industry is in private hands and open to foreign investors.

The beginnings of the shale gas industry date back to 2008 and the 13th Onshore Licensing round; a further 14th round was held in 2015. In that year, then Prime Minister David Cameron declared that the government was 'going all out for shale'. This statement came after a moratorium was lifted after a shale gas well at Preese Hall, drilled by Cuadrilla Resources, had triggered a minor seismic event, and drilling was banned while the regulations were revised. Since 2015 the national government, in various guises, has been supportive of shale gas development, maintaining that it is compatible with the UK's low-carbon energy transition, that it will improve energy security by reducing import dependence, and that it will attract significant investment and create numerous jobs where they are needed in the North of England. However, political support for shale gas is not universal. None of the devolved governments in the UK – in Wales, Scotland and Northern Ireland – is currently in favour of shale gas development, and there have been calls to ban hydraulic fracturing altogether. Thus, given the current Brexit-related turmoil in the UK, one would have to say that it is questionable whether there is long-term political support for the industry, particularly as the major opposition parties have vowed to ban it.

The UK has a nationwide high-pressure pipeline system – the national transmission system – and a comprehensive local gas distribution network. The UK gas price – reflected in the NBP – is the result of gas-on-gas competition and for a long time served as the EU's benchmark price, but was recently supplanted by the Dutch TTF price. The problem for shale gas investors is that almost all of the domestic industry is oriented to support offshore conventional oil and gas production. This means that there is limited domestic onshore capacity and current operations are dependent on US service companies. The UK government maintains that it has a 'Gold Standard' when it comes to shale gas regulation. The reality is that most of the regulations actually originate from the EU. Responsibility for regulation is shared between a number of national-level organizations – the Environmental Agency, the Health & Safety Executive, the Oil and Gas Authority, and the Department for Business, Energy & Industrial Strategy – while the local planning authority at the country level has to sign-off on planning permission at the site level. It is the local level that has proved to be the Achilles heel for the industry in England. Local politicians have responded to pressure from their constituents and environmental groups and have denied planning permission. This has resulted in court actions and public inquiries that have, for the most part, concluded in favour of the national government and the industry. However, this has significantly slowed the pace and added cost to the exploration process. In an attempt to court local communities, the UK government in 2014 announced an incentive scheme to distribute substantial financial benefits to local councils.[54] The industry was hoping to drill twenty to thirty exploration wells, and, so far, it has instead managed partially to fracture two wells. After a lengthy appeals process, in late 2018–19 Cuadrilla Resources ran into problems as its fracture operations triggered seismic events that were above the threshold imposed by a traffic-light system put in place after the initial problems in 2012.[55] The limit on permitted induced seismicity is very low by international standards and the industry is now saying that it cannot progress unless the limit is raised. A review was under way when further seismic events resulted in a new moratorium being imposed in late 2019.

The UK onshore oil and gas industry is the province of small companies such as Cuadrilla Resources, iGas and Third Energy, but recently it has been joined by the petrochemical company INEOS and its new business INEOS Shale. It is noteworthy that none of the oil majors has directly invested in UK shale, despite having long been involved in the North Sea and some being headquartered in the UK. Clearly, they see the reputational risk to be far greater than any potential commercial reward. The small companies have backers, including France's Total, who have been remarkably patient, but they may soon give up on the prospect of commercial shale gas production. The lack of progress is almost entirely due to highly effective campaigns by a coalition of local and environmental groups at various drilling sites. As elsewhere, the local protesters are concerned about the negative environmental impacts of shale gas development, while environmental groups maintain that shale gas development is incompatible with the UK's energy and climate objectives. Various attempts by the UK government to streamline the process and/or sideline local government have been unsuccessful. The UK government's own opinion poll survey, which has been running a question on shale gas since 2013, shows a gradual decline in support for shale gas, with over 30 per cent of respondents opposed to it by the end of 2018, compared to 21 per cent at the end of 2013.[56] However, over 40 per cent consistently reply that they neither support nor oppose it, principally citing a lack of understanding of the issue. Despite the relative indifference at the national scale, the industry does not seem to have a social licence to frack from the local communities in Northern England (there are similar disputes around onshore oil production in the South). In sum, in the UK case, the presence of a significant resource in place and support from the national government has not, so far, translated into a successful exploration phase, and the prospects for a UK shale gas industry anytime soon look rather dim. It is telling that the government's own assessments of future gas security do not assume future shale gas production.[57]

Shale gas with Chinese characteristics

To date, natural gas has played a modest role in China's national energy mix, accounting for 7.4 per cent of primary energy consumption in 2018. By comparison, coal accounted for 58.3 per cent, and on average, in the OECD, natural gas accounts for 26.6 per cent of primary energy consumption. As ever, given the scale of China, percentages can be misleading. In 2018, China produced 161.5 bcm of natural gas, accounting for 3.8 per cent of global production, ranking it sixth overall. However, the tide has clearly turned in China, which has embarked on a dash for gas as it seeks to reduce urban air pollution and reduce the carbon intensity of its economy. According to BP data, gas demand in China grew by 17.7 per cent from 240.4 bcm in 2017, to 283 in 2018. This accounted for 22 per cent of global gas consumption net growth.

In the EIA's estimates, China ranked first in the world with 31.6 tcm of technically recoverable shale gas resources. China's Ministry of Land Resources has published a lower resource estimate of 25.1 tcm.[58] Nonetheless, there would seem to be a substantial amount of shale gas in place in China; the problem is that the rocks are heavily faulted, the shale is deep (below 3,500 metres) and it is often found in mountainous, remote and arid areas (or areas where there is competition for water) that present significant logistical challenges. Despite these challenges, the Chinese government has made shale gas development a national priority, and in 2012 set a target of 100 bcm/year of combined tight gas, coal bed methane (CBM) and shale gas by 2020. In the face of slow progress, the shale gas target was subsequently reduced to 30 bcm/year, with a target of 80–100 bcm/year by 2030; nevertheless, there is clearly strong political support for shale gas development in China.[59] This has only strengthened as a result of the trade frictions with the US, and in late 2019 a research report from China's National Energy Administration suggested that the pace of gas demand growth would slow to around 10 per cent, as a result of slower economic growth and the pressures on gas infrastructure, and that efforts would be stepped up further to increase domestic output, including shale gas, tight gas and CBM.[60] The Chinese gas industry is dominated by the NOCs, particularly CNPC and Sinopec. These companies dominate domestic conventional production and control most of the long-distance pipeline networks. The best prospects for shale gas development are currently in the Sichuan basin, which is already home to conventional oil and gas production. The Chinese government did open up shale blocks to non-NOCs in 2011–12, but the companies that won licences do not have the experience or the financial clout to undertake a significant drilling programme. Consequently, CNPC and Sinopec are leading the industry in Sichuan.

At present, China's natural gas market needs reform, and significant investments in infrastructure. Natural gas prices are set by the central government, with some intervention at the provincial and city level. The system is complex and there is a delicate balance yet to be struck between setting a price that is low enough to encourage consumers to switch from coal to gas, but high enough to incentivize domestic production and investment in the required supporting infrastructure. As part of the so-called 'Blue Skies' policy, the government has set a target that the share of natural gas will reach 10 per cent in 2020 and 15 per cent by 2030. As detailed above, in recent years China has led the world in gas market growth, largely through increased LNG imports, and that growth seems set to continue as the Power of Siberia pipeline from Russia started to ramp up at the end of 2019, supplementing pipeline gas supplied from Myanmar and Central Asia. Initially, the plan to phase out coal from district heating and replace it with natural gas was so ambitious that construction of pipelines was not completed, resulting in gas shortages in the winter of 2017; but the completion of district heating networks has been made a priority, and Chinese companies strategically purchased larger amounts of natural gas prior to winter the next year.

The central government has provided subsidies to support shale gas production, which have now been extended to tight gas, and by late 2013 Sinopec and CNPC had already spent over $1 billion to develop shale gas, and will spend at least the same again, if not more, by 2020. There are accounts that, at the local level, a lack of regulation and enforcement have provided the NOCs with a free hand, creating problems for the local farming communities in the Sichuan basin. There have also been some significant seismic events, resulting in deaths and damage to property. The NOCs are enormous undertakings that have long experience of working with China's geology and operating conditions. These companies can also finance their own drilling programmes; thus, they probably represent the best bet for significant shale gas production.[61] They also control all aspects of the supply chain, and the Chinese industry has been developing its own shale gas equipment suited to local conditions. Various international companies have partnered with the NOCs, but this has proved unsuccessful as their imported experience failed to deliver in Chinese conditions. Despite the challenges, by 2017 some 1,000 wells of all sorts had been drilled in Sichuan and the surrounding area, with production growing from 4.5 bcm in 2015 to 7.9 bcm in 2016, with Sinopec's operations at Chongqing leading the way.[62] A study by Wood Mackenzie in early 2018 noted that the industry had grown to nearly 600 producing wells and 9 bcm of production, and that commercial well costs had fallen by 25 per cent since 2014, suggesting a learning curve. The consultants projected production to almost double to 17 bcm by 2020, requiring nearly 700 wells.[63] This would be well short of the 30 bcm target set by the government. In 2019, the consultants revisited their analysis and reduced their overall outlook for China's domestic supply.[64] Production is still expected to double from 149 bcm in 2018 to 325 bcm in 2040, but this is 39 bcm lower than their earlier forecast. This is because 'the long-term growth of CBM and shale production looks to be challenging'. Although shale gas production reached 10 bcm in 2018 – 7 per cent of domestic production – Wood Mackenzie think it will be difficult to duplicate Sinopec's success elsewhere. Consequently, they have downgraded their 2040 shale gas forecast to 88 bcm, 44 bcm lower than their 2017 view. While the environmental movement is not strong in China, Western media reports have suggested that there is growing local dissatisfaction with the environmental impacts of shale gas exploration and production. There is an obvious irony here: the central government's campaign to increase the role of gas and promote domestic production to improve the urban environment is coming at the expense of the quality of life in rural areas where shale gas is being developed.[65]

China's shale gas experience suggests that the country is developing the resource in its own way. So far, progress has been slow, conditions are challenging, and costs remain high. While it would be wrong to write off a shale gas revolution with Chinese characteristics, it seems highly unlikely that the scale of production will meet the ambitious targets set by the government. If demand continues to grow, and if domestic production disappoints, China will need to build more import pipelines and LNG terminals, and it will exert a strong influence on global gas prices and investment decisions.

Conclusions

It is hard to overstate just how dramatically the shale revolution has transformed the US energy landscape. In their 2019 *Statistical Review*, BP noted that: 'The growth rates in oil and gas output exceeded any other country's annual increase in our 50-year history.'[66] Furthermore, the report notes that the US holds less than 4 per cent of the world's oil and 6 per cent of gas reserves, but it now accounts for over 20 per cent of global gas supply and 16 per cent of global oil supply. As we have seen, a specific set of circumstances in the US has supported the shale revolution. These have proven difficult to replicate elsewhere. However, this was a revolution that took thirty years to evolve and it is still very early days for shale gas development elsewhere. In fact, we observe that modest growth of shale gas production has taken place in both China and Argentina.[67] Others may well follow, such as Australia where the initial focus has been on coal-seam gas. The experience in Poland suggests that the techniques that have been so transformative in North America may not work under different geological conditions, whilst the current plight of the UK shale gas industry demonstrates how 'above-ground' issues can confound the best intents of the national government to promote development. China is a different case again. Time will have to tell whether its state-owned oil and gas companies can overcome complex geological challenges and operating conditions to reach a level of production that will make a material difference to China's gas security. If successful, this would have significant implications for the global gas industry.

To date, the key consequence of the US shale revolution has been the dramatic turnaround of the supply outlook in the world's largest gas-consuming country. No longer is the United States concerned about security of natural gas supply. Instead, its primary concern centres around finding outlets and end users for its abundant natural gas. At home, abundant cheap gas is forcing coal out of the power-generation mix and providing a raw material for an industrial renaissance. The Trump doctrine (shorthand for the promotion of 'freedom gas', in particular as a counterweight to Russian natural gas in the EU) has been controversial from the start, and is likely mostly rhetoric, but an apt illustration of the altered posture of the US diplomatic corps following the rise in domestic production. Though we lack a counterfactual to prove the point, this new-found confidence in what President Trump calls 'American Energy Dominance' seems to play an important part in the country's belligerent stance towards the global trading system, and is enabling a more insular foreign policy based on sanctions and punitive tariffs, rather than military strength.[68]

Though it is tempting to make bold predictions about the future, we want to highlight that the US shale industry is young. Thus, at this point, we have more questions to grapple with than definitive answers. What we have learned is that the US shale industry has proven sceptics wrong on several occasions. Early analyses cautioned that US shale might only be short-lived; today we know not only that the reserve base is very significant, but also that very substantial amounts of natural gas can be produced at a low cost.[69] Add to this the record amounts of associated natural gas produced with tight oil and NGLs, and the notion that drilling practices and recovery rates are still subject to change,[70] and we can confidently conclude that the US will be producing a very competitive resource for decades to come, which both can meet domestic demand, and needs considerable export outlets.

That said, there are questions, too. As financiers turn their focus increasingly onto the profits and environmental performance of oil and gas companies (rather than expanding production), how will this affect the US natural gas landscape? With traditional domestic markets such as power generation increasingly saturated, can sufficient new demand be created, either domestically or internationally? Will public and political opposition to the fracking industry broaden in the years and decades ahead, and, if so, how does this impact the industry? By extension, how will natural gas be viewed and regulated in the era post the Trump Administration?

Time will provide us with data that can help answer these questions. For the time being, the emergence of shale is nothing short of a revolution. We now turn to a second revolution – in part spurred by, but also enabling, the continued growth of the US shale gas industry: the coming of age of LNG.

CHAPTER FOUR

The Coming of Age of LNG

Cooling natural gas to $-162°C$ ($-260°F$) produces LNG and reduces its volume to 1/600, which enables it to be transported by ship – as well as by truck and container – and can help with the development of reserves that are beyond the reach of gas pipeline networks. Equally, as is increasingly evident today, this enables countries that are beyond the reach of pipeline networks to tap into a market that had been the domain of a select group of countries for decades. Thus, the LNG industry plays a key role in making the gas market global.[1] Over the last sixty years, an established LNG supply chain business model has evolved that has put most of the risk of major upfront financial investments on producers, which was mitigated by cementing long-term bilateral relations in contracts, with limited flexibility to resell the product, or buy quantities other than previously agreed. In fact, one can think of the traditional model as a 'floating pipeline' linking a dedicated source of supply to specific markets that are tied to one another through strict contractual obligations (long-term contracts and destination clauses). However, fundamental changes are in process, specifically related to supply (as discussed in chapter 3), LNG technology, and shifting demand patterns that have challenged that traditional business model and that raise questions about the ongoing reconfiguration of the global LNG business, and where this may lead us. This chapter charts the evolution of the LNG industry and identifies the essential characteristics of the traditional model; it then turns to the impact of the Fukushima crisis – a key turning point in the evolution of the industry – and examines the developments since then that are challenging the fundamentals of the system. It concludes by considering the geopolitical consequences of the emergence of a more flexible and globalized LNG industry for both suppliers, consumers and new market entrants.

A brief history of LNG

The production of LNG was experimented with in the first half of the twentieth century, but it did not get off the ground, and was in fact slowed down following a major accident with a peak shaving plant in Cleveland in the US in 1944. About fifteen years later, high electricity prices and natural gas shortages incentivized the Chicago Union Stockyards, Continental Oil Company and British Gas Council to turn an old Second World War dry bulk carrier into an LNG ship, dubbed the *Methane Pioneer*.[2] In 1959, the modern LNG industry was born when the *Pioneer* shipped LNG from Lake Charles, Louisiana in the US, to Canvey Island in the UK. One of the motivations for the UK to import natural gas was to provide an alternative source of energy, cleaner than coal, which had produced the Great Smog of London in 1952.

After the discovery of the giant Hassi R'mel field in the Northern Sahara, Algerian state-owned oil and gas company Sonatrach became the first commercial exporter of LNG. In 1964, their GL4-Z (Camel) plant, with a capacity of 1.5 million tons a year (MT/y), became operational (it eventually closed in 2010). Algeria developed two LNG production centres: one in Arzew, and one in Skikda. Sonatrach is a 100 per cent owner of the liquefaction capacity in the country, and sold the overwhelming majority of LNG to clients in France, Italy,

Spain and Turkey. It is worth noting that Sonatrach is also a major supplier of European clients by pipeline (to Spain and Italy); however, this only came to fruition after the 1970s, at the expense of domestic consumption.[3] In recent years, various challenges regarding feed gas production, including rising domestic demand for natural gas (further exacerbated by generous energy subsidy regimes) and political risk (in the form, amongst other things, of excessive red tape) have continued to plague Algerian domestic production.

Following Algeria, more liquefaction plants were built in Alaska, Libya, Brunei, Abu Dhabi and Indonesia, serving clients in Northwest Europe, the US and Northeast Asia. A major share of the initial build-out of liquefaction capacity, starting in the 1970s and onward, was coordinated by Japanese companies, colloquially known as 'Japan Inc'.[4] Following the oil crises in the early 1970s, Japanese policymakers and companies decided to move away from oil as a primary energy source, and as a source for power generation. Nuclear energy and natural gas were the major winners in this transition. In the late 1980s, South Korea joined the LNG club, and in the early 1990s Taiwan did as well. For a period, these three nations dominated the LNG import market. Even as recently as ten years ago, these three countries accounted for 62 per cent of global LNG imports, with Japan by far the largest importer (92.12 bcm).[5] As discussed later, this picture has slowly but surely changed fundamentally, even though the energy strategies and fuel choices in these countries mean that they will continue to consume substantial amounts of LNG for many years to come.

The early decades of the LNG market were a relatively exclusive undertaking, with only a few countries exporting the fuel, and a similarly modest number importing it. A consequence of this is that the industry structure has long been fairly rigid.

Exporters have to make substantial upfront capital investments to construct liquefaction capacity, often with supporting infrastructure to develop gas reserves and bring the feed gas to the liquefaction plant. Because LNG prices have traditionally been linked to the oil price (an imperfect solution absent a global market for natural gas), doing so entailed a major price risk, which typically lay with the supplier. In order to finance such projects, supplier and buyer would negotiate long-term offtake agreements (in the order of magnitude of twenty years), often with so-called 'take-or-pay' volumes in them – in other words, minimum levels of natural gas that the offtaker would pay for, regardless of whether or not that company would in fact need the natural gas. In addition, there were restrictions (so-called destination clauses) that limited the buyer's ability to resell the natural gas to third parties. In other words, buyers in this cooperation typically carried a major volume risk. Slowly but surely, the LNG group expanded, new sources of LNG came online and new importing countries emerged. According to the LNG importers' association (GIIGNL – the International Group of Liquefied Natural Gas Importers), there are now forty-three countries importing LNG, compared to eighteen ten years ago.[6] In 2008, BP's *Statistical Review* reported LNG exports from fifteen countries (excluding re-export from Belgium), and production totalled 226.51 bcm. Roll forward to 2018 and BP reports eighteen exporting countries and total production of 431.0 bcm.[7]

The making of a global market

Trade in LNG – and with it, arguably, the globalization of natural gas markets – started in the 1960s, but did not truly spread until much later, and it took a country of barely 12,000 km^2 in size to help the industry do it.[8] Prior to the age of hydrocarbons, the Qatari region (Qatar became an independent country in 1971) had mostly made money through fisheries and pearl diving. This changed in 1940, when oil was discovered, and this became the chief economic activity in the emirate. In search for more oil, in 1971 Shell discovered the North Field off Qatari shores, which later proved to be the world's largest non-associated natural gas field. At the time this was considered a disappointing find, both for the IOC and for the Qatari leadership, since natural gas had only recently emerged as a marketable product in most of the world. Moreover, there were uncertainties about the ownership of the field, since a part of it crossed into the exclusive economic zone of Iran. Eventually, Shell handed the concession back to Qatar, after deciding to focus new exploration activities elsewhere.[9]

As a result of the oil crises in the 1970s, the revenue stream from Qatar's (by regional standards) modest oil reserves was still significant, but in the 1980s oil revenues started to decline, and new exploration activities were largely unsuccessful. The IOCs were losing interest in the country. The leadership in Qatar thus started thinking about a strategy to turn natural gas and liquids into a revenue stream. First, there came a heavy focus on using natural gas in parts of the domestic energy economy, specifically for electricity generation, water desalination and industrial consumption (petrochemicals, and fertilizer production). Second, Qatar started developing plans for exports, both seaborne trade and, controversially, regional pipeline trade. For a variety of reasons – mainly geopolitical, rather than economic – most of the pipeline plans did not materialize.[10] The Kingdom of Saudi Arabia found its own natural gas deposits, and plans to cooperate with Qatar were shelved. Moreover, the Kingdom's rulers decided not to allow Qatar access to its exclusive economic zone, making the construction of a pipeline to nearby Kuwait impossible (today Kuwait

imports LNG). Bahrain's natural gas consumption (and outlook) were very modest, and the country could not resolve a
border dispute with Qatar, and trade between these two sides
did not materialize either. Though the intricacies of this project are beyond the scope of this chapter, it is worth noting
that eventually a pipeline between Qatar and the United Arab
Emirates (Abu Dhabi) did materialize: the Dolphin pipeline,
which started operations in 1999, and brings natural gas to the
Emirates, and Oman.

Realizing Qatar's natural gas strategy proved challenging for
several reasons. The country lacked the institutional decision-
making capacity to get complex capital-intensive projects off
the ground. To give an example, all cheques worth more than
US$50,000 had to be signed by the Emir personally.[11] In addi-
tion, Qatar Petroleum at the time lacked the relevant expertise
to move projects forward unilaterally. Several IOCs – namely
Total, British Petroleum and Shell – initially were interested
in participating. However, in 1983, Shell left, as it focused its
attention on other parts of the world, notably Australia. The
others stayed, and together established Qatargas in 1984,
and started searching for possible buyers for Qatar's LNG.
Conversations with several major Japanese companies were
fruitful (following the oil crises, these companies were looking
to reduce their imports of crude oil, and increasingly stringent
domestic air quality rules made natural gas an obvious alternative), but geopolitical turmoil complicated progress. The war
between Iraq and Iran, and specifically the bombing of several
tankers from neighbouring countries, cast a shadow on shipping trade. To Doha, these developments also confirmed that
the Kingdom of Saudi Arabia and/or the Gulf Cooperation
Council were no safeguards against violence in the region, and
that making friends in other parts of the world might be a good
preservation strategy. As a direct consequence of the turmoil,

Japanese firms made investments in Australia instead, and Qatar had to wait for the next wave of long-term contracts to be signed.[12] This reflects the 'lumpy' and highly cyclical nature of LNG supply growth.

In September 1991, twenty years after Qatar's independence, natural gas production from the North Field started. The resource was used for domestic consumption, and it enhanced oil recovery. Conversations about exports in the form of LNG continued, but not without challenges. British Petroleum left Qatar, but American Mobil joined. Together with several major Japanese companies – including the utility Chubu and a number of Japanese financial institutions – eventually agreements were reached, and investments made, and the first train started operations in 1996. In 1997, the first exports to the Republic of South Korea followed. Construction and operation of the various plants in the country were done by two project companies: Qatargas and RasGas. Under Emir Tamim bin Hamad Al Thani (the previous emir had been deposed by his son in a bloodless coup in 1995), natural gas exports received full state support, and by 2011 the nameplate capacity of a series of LNG trains had reached 77 MT, making Qatar the world leader of LNG by a landslide. In 2012, Qatari LNG exports totalled 105.5 bcm (78 MT), which accounted for 28.3 per cent of global exports.[13] Whilst the initial focus of the Qatari companies had been east of Suez (the Pacific basin), once LNG market prospects west of Suez (the Atlantic basin) rose, both Qatargas and RasGas started to search for sales there as well.[14]

The UK's natural gas production from the North Sea peaked in 2000, and in 2004 it became a net importer of natural gas. In anticipation of decline, the natural gas industry began to invest in new import infrastructure. First, a pipeline (Frigg) from the Norwegian sector of the North Sea, and then interconnectors to the European continental gas market via Belgium and the Netherlands. It also involved the construction of three LNG import terminals: first the Grain terminal in Kent that is owned by National Grid; then two terminals were built at Milford Haven in West Wales. The largest of these is South Hook, which is part of an integrated supply chain that brings Qatari LNG to the UK. The shareholders in the terminal are Qatar Petroleum, ExxonMobil and Total. The South Hook terminal has provided an important outlet for Qatari LNG and, until recently, Qatar accounted for more than 90 per cent of the UK's LNG imports. Most recently, new supplies from Russia and the United States have helped to diversify the UK's LNG imports, with Qatar only accounting for 55 per cent of imports in 2018. The third import terminal, Dragon LNG, was until recently a joint venture between Petronas and Shell (previously the BG Group). In July 2019, Petronas sold its 50 per cent share to Ancala Partners, an independent infrastructure investment manager, but it will continue to import LNG via the facility. This build-up of LNG import capacity – with a total send-out capacity of 48.1 bcma – in the UK was paralleled by similar development in maritime Europe, whereas other countries invested in regasification capacity to diversify their import portfolio. Thus, the rapid expansion of Qatar's LNG production provided linkage between the gas markets in Europe and Asia. In May 2019, Qatar, through a Luxembourg company, announced investments in gas-fired power-generation capacity in Belgium, to replace the planned phase-out of nuclear plants in that country in the years ahead.[15] This illustrates that companies are increasingly willing to make downstream investments to secure longer-term demand for natural gas, as competition in the LNG market intensifies.

In 2005, after several anomalous drilling results, Qatar announced a self-imposed moratorium on further development

Table 4.1 Total nameplate LNG liquefaction capacity in Qatar, as of December 2018

Liquefaction unit	Year of start-up	Number of trains	Capacity MT/y
Qatargas 1	1997	3	9.5
Ras Laffan 1	1999	2	6.6
Ras Laffan train 2	2004	1	4.7
Ras Laffan 2 train 2	2005	1	4.7
Ras Laffan 2 train 3	2007	1	4.7
Qatargas 2 train 1	2009	1	7.8
Qatargas 2 train2	2009	1	7.8
Ras Laffan 3 train 1	2009	1	7.8
Qatargas 3	2010	1	7.8
Ras Laffan 3 train 2	2010	1	7.8
Qatargas 4	2011	1	7.8
Total current capacity		**14**	**77.0**

Source: available at www.qatargas.com/english/operations/lng-trains.

of the North Field to rule out uncertainties regarding long-term production. This moratorium was renewed several times, and then finally lifted in April 2017.[16] The initial announcement suggested a further planned expansion of liquefaction capacity by three trains, plus debottlenecking the existing trains, raising nameplate capacity to 100 MT. Later these plans were adjusted, and Qatar now plans to add four new liquefaction and purification facilities to its existing capacity, expanding total export capacity from 77 to 126 MT by the middle of the 2020s. In doing so, Qatar has signalled to other exporters, both existing and new, that the country intends to continue to play a prominent role in the market for LNG. As we discuss in more detail in the next section, the lifting of Qatar's moratorium comes as project developers in other parts of the world are also looking to bring new projects to a final investment decision (FID), and demand for LNG is booming in

comparison to other fossil fuels. Being a known low-cost producer of LNG, because of the significant amount of condensate
and liquids that are produced in Qatar that cross-subsidize
LNG production, the announcement should probably also
be interpreted as a deterrent to the higher-cost producers of
LNG seeking to reach an FID. By late 2019, we can conclude
that, if Qatar's announcement was to deter new investments,
this strategy did not work out. Starting in 2018, a series of
new liquefaction projects has taken FID, in places as diverse as
Canada, the United States, Mauritania/Senegal, Mozambique
and Russia. In their *Global Gas Security Review 2019*, the
IEA reported that a record amount of new liquefaction capacity would be sanctioned in 2019, adding a further 170 bcm of
new production, which far surpasses the previous record of 70
bcm set in 2005.[17]

In addition, the timing of the announcement to lift the moratorium should be seen in its geopolitical context, which is
one of a deep rift between Qatar on the one hand, and the
Kingdom of Saudi Arabia, Bahrain, United Arab Emirates and
Egypt on the other. It is worth noting that, despite the block-
ade that the coalition implemented regarding Qatar, pipeline
natural gas trade with the Emirates has not been affected, and
Qatar has expressed no intention of stopping the flow of gas
in the future.[18] Regional tensions aside, the rapid development
of Qatar helped to propel LNG trade into a truly global business, and build a prosperous state on the revenues. Geography
also plays a key role in terms of Qatar's significance to the
global LNG market. It is a swing producer, in the sense that it
can supply clients in both the Pacific and Atlantic basins. The
majority of its exports go to Asia, but, as noted above in the
context of the UK, when Asian demand is soft, natural gas can
flow elsewhere too. We anticipate that in the future we will see
Qatari LNG pursue this option more often.

Next to occasional shifts in global demand and supply, an additional factor to consider is the expiration of existing offtake agreements. In the decade ahead, a significant number of existing contracts (for an estimated 100 MT) with buyers in both the EU and East Asia are going to expire. These supplies mostly originate from Qatar, Malaysia, Algeria and Nigeria. In many cases (assuming continued availability of natural gas), suppliers will likely seek to renew these offtake agreements, and some likely will. However, given the global availability of supplies from elsewhere, and growing competition, many of these contracts may not be renewed. In addition, in light of structural changes in the market, LNG contracts will likely look different as well.[19] There is evidence to suggest that contract period, flexibility of the start date of the contract, flexibility on annual delivered volumes, take-or-pay clauses, the right to divert cargos, credit security and transfer of title (ownership) are all subject to change at this point, generally resulting in more risk for sellers than has traditionally been the case.[20] Given that the liquefaction capacity of projects in countries such as Qatar, Algeria and Malaysia have now been depreciated, companies will be able to sell cargos of LNG at very competitive prices, if desired. Thus, Qatar has been able to maximize the revenues obtained from its LNG trade and it can also benefit from arbitrage opportunities, should they exist, between the Asia and Atlantic basins. However, as more supplies from North America, Russia and East Africa come to market, these will impact historical trade patterns.

As we discuss next, the formation of a truly global market takes time, and is arguably still a work in progress. What helped next was the major expansion of the global proven reserve base, with the advent of shale gas and CBM (or CSG), on the one hand, and a shift of global demand growth to countries that often lack pipeline connectivity, on the other hand. In addition, as we discuss below, singular events can help propel a market in a certain direction.

Coming of age takes time

On 11 March 2011, the Tohoku earthquake off the coast of Japan was followed by a devastating tsunami. Even though the active nuclear reactors on the coastline, including those in Fukushima, immediately shut down, the tsunami destroyed the emergency generators cooling the reactors at Fukushima, causing three reactor cores to overheat and subsequently to melt down. This, in turn, led to the release of radioactive material, and the shutdown of Japan's nuclear fleet as a precaution. Consequently, Japanese utilities, in order to safeguard electricity supply, had to turn to the market and scramble for LNG.[21] The short-term implication was a spike in the price for LNG, specifically in the Pacific Basin, but there have proved to be profound long-term consequences as well. As prices on the spot market rose, this sent a message to natural gas producers around the world that there were profits to be made from trading natural gas in Northeast Asia. As the LNG price in Asia is indexed to the price of crude oil, the recovery of the oil price after the global financial crisis of 2008 also had some bearing on the attitudes of investors. At the time, conventional wisdom held that $100 a barrel of oil was the 'new normal'.[22] Figure 4.1 compares the crude oil price with Japan's LNG price, the UK's NBP – at the time the most important benchmark for Europe – and the US Henry Hub. The immediate aftermath of Fukushima saw a great divergence between the European and Japanese gas prices – consequently, most LNG went to Asian markets. The low price of the Henry Hub, relative to the Asian market, added further impetus to plans to develop a US LNG export industry.[23]

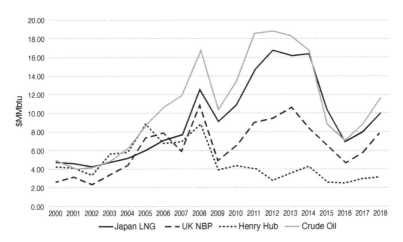

Figure 4.1   Natural gas prices, 2000–2018

Source: BP (2019), *BP Statistical Review of World Energy, June 2019.* London: BP, p. 37.

In the period prior to 2011, natural gas production in the United States and Australia had grown substantially with the rise of shale gas and CBM (see chapter 3). As domestic markets in these countries were increasingly saturated and, particularly in the United States, production kept growing, producers were eyeing other outlets to sell their product. Thus, in the aftermath of 'Fukushima', a significant number of liquefaction projects reached their FID. These projects, in turn, have been coming to market since, the last ones starting operations in 2020 at the latest. It is important to note that, prior to the disaster in Fukushima, a number of liquefaction projects, specifically in Australia and Papua New Guinea, had already taken their FID, based on more general factors such as anticipated GDP and energy demand growth.

Table 4.2 provides the details of all projects that have taken their financial investment decision in or after 2012. This

Country	Name	Start date	Nameplate capacity (MTPA)	Lead company*
\multicolumn{5}{l}{Table 4.2 LNG projects entering into production between 2016 and early 2019}
Australia	GLNG T1	2016	3.9	Santos
Australia	Australia Pacific LNG T1	2016	4.5	ConocoPhillips
United States	Sabine Pass T1	2016	4.5	Cheniere
Australia	GLNG T2	2016	3.9	Santos
United States	Sabine Pass T2	2016	4.5	Cheniere
Australia	Gorgon LNG T1	2016	5.2	Chevron
Australia	Gorgon LNG T2	2016	5.2	Chevron
Malaysia	MLNG T9	2017	3.6	Petronas
Australia	Australia Pacific LNG T2	2017	4.5	ConocoPhillips
United States	Sabine Pass T3	2017	4.5	Cheniere
Australia	Gorgon LNG T3	2017	5.2	Chevron
United States	Sabine Pass T4	2017	4.5	Cheniere
Australia	Wheatsone LNG T1	2018	4.45	Chevron
Russia	Yamal LNG T1	2018	5.5	Novatek
United States	Cove Point LNG	2018	5.25	Dominion
Cameroon	Kiribi FLNG	2018	2.4	Golar LNG
Australia	Wheatsone LNG T2	2018	4.45	Chevron
Russia	Yamal LNG T2	2018	5.5	Novatek
Australia	Icthys LNG T1	2018	4.45	INPEX
Russia	Yamal T3	2019	5.5	Novatek

* Company with largest stake in the project.

Source: International Gas Union (IGU) (2019), *World LNG Report 2019.* Barcelona: IGU, p. 98.

represents a total additional capacity of 91.5 MT. In 2012, BP's *Statistical Review* reported total LNG production as being 327.9 MT; by 2018 that total had increased to 431.0, but it often takes time for new plants to meet their nameplate capacity and some existing plants went out of production or were subject to maintenance. Nevertheless, this represented a 20 per cent growth in annual production, and there are still several projects to come on line. The IGU's list of LNG projects currently under construction or sanctioned suggests an additional 47.25 MT in 2019 and a further 15.5 MT in 2020; the bulk of this investment wave is in the US, with the few remaining Australian projects also coming on line. Thus, by 2020, the industry will have added 225.45 MT of new liquefaction since the Fukushima crisis.

In 2016, the anticipated growth in liquefaction capacity led to a widespread belief that a 'glut' of LNG was upon the world. With supplies from the United States and Australia scheduled to come to market, prices were going to drop, and the higher-cost projects were anticipated to go 'underwater', possibly for a considerable amount of time. This did not happen until 2019, due to a combination of factors. First, the projects did not all come to market as planned: there were delays in both Australia and the US, spreading their impact on available global supply out over the years. Second, demand responded. With economic recovery after the 2008 crisis, energy demand rebounded, and with it demand for fossil fuels. Natural gas has become the fastest-growing fossil fuel in all projections, a trend that is anticipated to continue (this is discussed in detail in chapter 5). Consequently, and in combination with rising coal prices and a more robust carbon price under the EU Emissions Trading Scheme (ETS), natural gas demand also recovered in the EU, a market long believed to have seen its last growth spurt. Going forward, there is room for continued growth, incentivized by the planned/regulated phase-out of both nuclear and coal-fired plants in the power sector, though regulatory uncertainty may cloud that outlook.[24] More importantly, and for two reasons, natural gas experienced significant growth in emerging economies. First, in countries such as China, air pollution has reached a level that policymakers no longer deem tolerable. In response, aggressive clean air targets have been adopted in the so-called 'blue skies' policy. In short, the central government orchestrated the rapid phase-out of coal from district and industrial heating systems in major urban areas, and banned the most polluting forms of transportation. As noted in chapter 3, the government-orchestrated goals to phase out coal from the industrial and space-heating sectors in 2017 were so ambitious that distribution companies were not able to build-out all the required natural gas infrastructure in time, leading to localized shortages of natural gas in the middle of winter.[25] It was commonplace for LNG to be moved by trucks in the absence of a local pipeline distribution network. In turn, clean air policies for 2018 were slightly relaxed because the Chinese government wants to avoid people sitting in the cold, but the medium-term outlook for natural gas in China has not changed: the most polluting fuels, particularly coal and some oil products, will face increased regulatory scrutiny, and renewables, nuclear energy and natural gas are to gain.

It is likely that, in due course, more emerging economies in East and South Asia will follow a similar trajectory – the extent of this chiefly depending on the competitiveness of LNG. In China, local air pollution was tolerated for a long time, and rigorous action only designed and implemented after air quality – or the lack thereof – became a public talking point, and studies started to demonstrate that life expectancy in urban areas in Northeastern China was 5.5 years lower than it would have been if the air were cleaner.[26] More recent data suggest that air pollution levels in other countries – for instance in India, Indonesia and Pakistan – have reached comparable, and sometimes even worse, levels than China has had for a long time. With air quality in Chinese cities improving, in 2019 nine of the top ten most polluted cities in the world are in India.[27] In February 2018, the Indian government announced plans to address air pollution drastically in the short and medium term.[28] However, contrary to China, not all emerging economies have the financial clout to support a policy-driven shift away from more polluting fuels to natural gas. In China, where city-gate prices for natural gas continue to be regulated, natural gas was not always price-competitive, meaning that someone in the supply chain (generally the NOCs) has to absorb the additional cost. One may still argue that this is a desirable policy to support the public good of cleaner air, but someone has to pay the full cost. In some emerging economies, this may be more complicated to resolve, and demand may therefore be more price-sensitive than seems to have been the case of China.

Given the size of its energy demand, it is understandable that most analysts tend to focus on countries such as China when forecasting natural gas demand. However, the implicit risk of overly focusing on a couple of large countries is that other relevant developments may stay below the radar. The growth in natural gas demand in recent years in smaller emerging economies, which helped absorb the build-up of LNG supplies, is a case in point. Analysts who wrote about the impending 'LNG glut' simply compared the demand outlook in major countries with the investment decisions that were made in the early 2010s, and concluded that the market was not going to consume all of those additional resources. In reality, one major contributor to proving that this was a false assumption was the aggregate demand growth in emerging economies, often using floating storage and regasification units (FSRUs) to start importing LNG. The latter is a relatively novel technology, first put in operation in 2004 by Excelerate, and can be either purpose-built, or a redesigned old LNG vessel.[29] By design, these units are much smaller than traditional onshore regasification infrastructure, so not only are they more flexible, but also they are significantly less costly, and therefore easier to finance. They can also be relocated, should an importing country no longer desire to import LNG. Slowly but surely, countries around the world have started to adopt this technology, by either purchasing or leasing FSRUs, and developing commercial models around them. Relatively modest LNG (and oil) prices further incentivized countries to turn to natural gas as a fuel source. Consequently, between 2015 and 2016, the aggregate demand of a sample of ten non-usual suspects when it comes to natural gas consumption was in fact, for some time, larger than that of China.[30] To be fair, structural challenges regarding gas demand growth in emerging economies persist, and in 2018 FSRU-based LNG demand declined a little. In fact, some of the most promising FSRU projects are now in developed economies, such as Australia, Ireland and Croatia. Nonetheless, FSRU technology has helped to democratize LNG, and with an occasional nudge by development banks will most likely help to expand further the pool of countries that can start buying the fuel.[31]

In 2019, it seemed that LNG supply had finally outpaced demand despite continued growth, resulting in lower prices for spot cargos, and more pressure on traditional oil-linked contracts. However, the concern is that, after 2020, the paucity of new FIDs in the last few years could result in a tighter market in the mid-2020s as, by then, demand growth could have absorbed all that new production. Much depends on whether or not China's appetite for natural gas will be translated into firm LNG demand.[32] In this context, the announcement of FIDs at various new liquefaction projects starting in 2018 have been heralded by industry analysts as the first of the next wave of LNG projects aimed at meeting demand in the mid-2020s, but, given the record level of FID activity in 2019, there are now concerns about a new supply glut in the late 2020s.

The exclusive buyers' club that we described earlier in this chapter, in 2018 was no longer. At the time of writing, forty-three countries have installed regasification capacity that can be used if desired.[33] Some of these new importers, such as Poland, have built a traditional onshore regasification terminal. Many new entrants, however, are using FSRU technology. In September 2019, twenty-four FSRU vessels were in operation, with another nine under construction / ordered. The latest innovation is the inception of so-called mini-FSRUs, which have a storage and send-out capacity that is sufficient to generate 200–250 MW of electricity and is ideal for island communities who wish to switch from oil to cleaner natural gas-fired power.[34] There are various reasons why countries may consider joining the LNG market. Early adopters of FSRUs, such as Argentina, chiefly wanted an option to counter anticipated supply shortfalls in domestic production. However, with the development of its shale gas potential, Argentina is now exporting LNG using a floating LNG (FLNG) vessel. Notwithstanding specific occasions when companies in gas-producing countries cannot get infrastructure built (New South Wales, Australia, or New England, United States) this will likely make regasification capacity obsolete. In Egypt, mismanagement of the hydrocarbon sector and a lack of exploration activities led to a sharp drop in domestic production, and, as consumption continued to rise, an FSRU offered temporary relief as Egypt became a major importer of LNG in 2015 and 2016. The Sisi government prioritized domestic exploration again, and ENI discovered and developed the giant Zohr field offshore of Egypt's coast, turning the country's fortunes once more. In October 2018, the FSRU lease was ended.[35] The net result is that Egypt will likely go from being an exporter of natural gas to an importer and back to an exporter within the space of ten years. Brazil's objective for leasing an FSRU was access to the waterborne gas market to be able to use natural gas for power generation in case of falling hydroelectric production (due to a lack of rainfall). In several Gulf Arab countries, such as Kuwait and the UAE, import facilities have been leased for a longer period to help meet rapidly growing domestic demand (these countries have significant domestic resources, but they have a high hydrogen-sulphide content, and are therefore expensive to develop). Bahrain is considering joining the market as well. Other countries, such as Lithuania, have used the lease of an FSRU to diversify the local supply mix, and to negotiate better terms under an existing long-term contract with Russia's Gazprom. Croatia is constructing regasification capacity for a similar purpose. For emerging economies, FSRUs have provided an option to tap into the LNG market to fuel industrial activity, or power generation. Pakistan, and more recently Bangladesh, are good examples of this. Both countries used to have significant domestic production, and domestic infrastructure to support its consumption. However, with production dwindling, imports have been on the rise, but credit challenges often stood in the way of investing billions of dollars in traditional regasification terminals, or paying for LNG. Smaller-scale solutions have been easier to finance, in cooperation with development banks (some of whom favour supporting natural gas rather than coal), and as a result existing domestic infrastructure has been used again. For remote locations, or island states, constructing pipeline infrastructure may not be economical or technically feasible. FSRUs have enabled these countries or regions to import natural gas, nonetheless. These can be developing countries, such as the Philippines or Indonesia, but also mature economies that face challenges in building-out pipeline infrastructure, such as Ireland (where post-Brexit security of supply considerations may help inform policy decisions) or Australia (where, despite very significant domestic production in the Northwest and

Northeast of the country, building infrastructure to major markets in the Southeast has proven to be complicated). Finally, smaller-scale FSRUs are enabling countries of modest size, e.g. Jamaica, to turn to natural gas as a cleaner fuel – as opposed to fuel oil – for power generation. In short, there are many different reasons why the group of LNG importers has expanded so dramatically over the last two decades, and it is fair to suggest that this expansion will continue in the years ahead, but LNG must remain affordable as many of these new importers are not as creditworthy as traditional importing countries.[36]

Two factors have enabled this development and are necessary for its continuation: first is the availability of natural gas, which at present means the continued development of new LNG capacity in North America and elsewhere. Second is ever-lower entry barriers into the market for seaborne gas trade, both from a technology standpoint, and also in terms of the flexibility of trade, which includes the ability to sign shorter-term contracts, easier access to capital, and the possibility to trade on an increasingly liquid spot market. Flexibility of trade has increased as a result of two developments: increased sales of LNG on a 'free-on-board' (FOB) basis, and legal measures prohibiting so-called 'destination restrictions'. The former is a relatively new concept in the LNG world, introduced in 2012 with Trinidad and Tobago LNG, and applied more broadly by several US companies starting in 2016. Historically, cargos have generally been delivered ex ship (DES), meaning that the seller was responsible for the delivery of a cargo to an agreed port of arrival (destination). The buyer in turn signed long-term offtake agreements, in doing so providing the seller with certainty that agreed volumes would in fact be purchased. A conventional LNG plant runs at maximum capacity, only shutting down for routine maintenance, and this means that supply is relatively inflexible and cannot be 'turned down' when demand falls. Because the market, for a long time, was relatively small, buyers and sellers were forced to work closely together, and contracts reflected that, in case of changes in the market, parties had relatively limited other options. Therefore, ships were often designed for a specific purpose – or even trade route – and regasification terminals and arrival ports were constructed with specific ships, and natural gas quality, in mind.[37] There are also differences in the composition of the LNG supplied from different sources, which raises issues of interchangeability (captured by the Wobbe index), as well as differences in the physical infrastructure used in different ports.[38] The former can involve additional treatment costs to enable LNG to be used in domestic pipeline systems, while the latter is the result of particular LNG projects supplying particular ports. A negative consequence of the various restrictions on resales of LNG cargos was that short-term trade had long been a rare phenomenon in the LNG sector. In Europe, the European Commission was increasingly of the view that such restrictions were not compatible with the operation of the single market, but LNG still played a modest role in the EU's import strategy; that has changed in recent years and flexible LNG is now an important part of the EU's gas security strategy. In July 2017, the Japan Fair Trade Commission (JFTC) ruled that destination restrictions in FOB contracts are illegal – and, under certain conditions, in DES contracts as well.[39] This was perhaps not the firmest ruling that the JFTC could have given. However, in combination with a growing sense amongst buyers that, because of the uncertainty of Japanese demand resulting from the controversy of bringing nuclear plants back into operation, flexibility is important for effectively managing your portfolio, it is likely that destination restrictions in Japan will also slowly disappear. Evidently, the ample availability of supply helped to change fundamentally the mindset of buyers.

The growth in opportunities for trade also becomes clear when one considers the new players that have entered the market in recent years. One evident example is the growing role of trading houses. Collectively, Trafigura, Gunvor, Vital and Glencore in 2017 traded an estimated 27 MT of LNG, or 9 per cent of total traded volumes.[40] In addition, several major IOCs – the so-called aggregators – have built up a portfolio of LNG supplies, in various locations and under various conditions, to take advantage of price differences between various parts of the world. Shell, Total and BP are examples. But trade is not just spurred from the supply side. In recent years, major LNG buyers such as Korea's Kogas or Japan's Mitsubishi have taken equity positions in upstream projects that give them privileged access to production. These companies in turn are themselves responsible for finding an end user for their natural gas, either in their domestic market, or elsewhere. By extension, various major LNG buyers have announced cooperation agreements with peers in other parts of the world, the aim being to share risk, but also to optimize portfolios, and reduce costs. In 2016, for example, the British utility Centrica announced such a cooperation with a Japanese natural gas utility, Tokyo Gas.[41] All these developments will contribute to the formation of an LNG market that is becoming more flexible, and more global.

Despite the growth of natural gas demand, it is important to note that most analysts still anticipate a relatively soft market in the short term, until 2020/1, as more supplies are scheduled to come to market. In 2019, spot prices were at historical lows as a slowdown in demand growth, falling oil prices and growing supply all combined. Low spot prices for LNG will put more pressure on sellers that have linked their LNG to oil prices, as buyers may be tempted to renegotiate their existing contracts. This reflects the cyclical nature of the LNG market, whereby, depending on your demand assumptions, at times supply may outstrip demand (having a downward effect on prices), and vice versa. Nonetheless, there is no shortage of project developers that are trying to move new liquefaction projects forward, and, at the time of writing in late 2019, it is clear that by the middle of the 2020s we will see another wave of new capacity coming to market. An emerging challenge for some companies is that, in a relatively well-supplied market, some buyers have appeared to be reluctant to sign new offtake agreements, as they feel more comfortable that they can source the natural gas they need. Increasingly mercantilist politics – e.g. the trade dispute between the United States and China – also play a role in this. The challenge then is that most project developers still need long-term offtake agreement to underpin their loans, in order to finance their projects. Some companies have pockets deep enough to finance projects off their balance sheet, as happened in October 2018 with LNG Canada, when Shell announced its FID, together with a group of Malaysian, Japanese, Korean and Chinese equity shareholders.[42] BP did the same with its project in Mauritania/Senegal, as did Qatar Petroleum and Exxon Mobil in early 2019 with their Golden Pass project. Yet for other companies this is not an option, and absent long-term contracts, or innovative thinking in the finance community, some companies may struggle to reach their FID. In 2018 though, we witnessed a rebound of long-term offtake agreements being signed (see figure 4.2), though with greater pricing flexibility than was the case in the past, and so for the moment we can observe the emergence of a multiple-tier market, in which some of the majors finance new liquefaction projects off their balance sheets, trusting that they and their equity partners will be able to find end users for their LNG, while other project developers are able to sign long-term offtake agreements to help finance their projects and move forward.

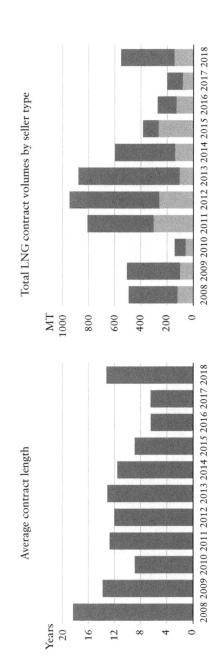

Figure 4.2   Average contract length and total LNG contract volumes by seller type

Source: reproduced with permission from Royal Dutch Shell (2019), *Shell LNG Outlook 2019.* Available at www.shell. com/promos/overview-shell-lng-2019/_jcr_content.stream/1551087443922/1f9dc66cfc0e3083b3fe3d07864b2d0703a25fc4/lng-outlook-feb25.pdf.

As described, the LNG market has changed markedly over the last twenty years (not least in comparison to the pace of change in the forty years before). Yet maturing of a market does take time, and current challenges in the LNG market demonstrate that it is no exception to this rule. First, an evident long-term challenge is the slow pace of market reform in large parts of the world, importantly including countries where natural gas demand is growing the most. The lack of third-party access to regasification terminals in Japan (which disincentivizes newcomers to enter the market and compete), or the dominance of Chinese state-owned companies on their domestic market, are important examples. In turn, this lack of a market-driven habitat has consequences for natural gas price formation in East Asia. Despite various initiatives to develop a price benchmark based on a trading hub, as in North America or Northwestern Europe, and despite the arguable major growth in short-term trade in East Asian markets, there is reason to be sceptical that an Asian LNG price based on market fundamentals will emerge soon.[43] In many MENA countries, natural gas price subsidies are still commonplace, making the import of natural gas in the form of LNG an activity that seems economically unsustainable in the long run. Second, infrastructural challenges continue to complicate natural gas reaching end-use markets – as is, for instance, the case in large parts of India (amongst other challenges).[44] In countries with a natural gas history and, therefore, infrastructure in place, such as Pakistan and Bangladesh, 'reconnecting' to the natural gas market is relatively easier than for countries that have to build from scratch. Third, as the market is developing a highly cyclical nature, at times supplies may outstrip demand, and vice versa. A period of high prices in the early 2020s might lure new entrants to the market and at the same time dampen demand, laying the groundwork for a market with abundant supply and relatively low prices in the second half of the decade. Finally, and an ongoing challenge for the gas industry more generally, there are uncertainties about the competitiveness of natural gas *vis-à-vis* other fuel sources – for example coal in the electricity sector – surely, in emerging economies, this is a major point of focus.

A bright near-term outlook for LNG

Notwithstanding the challenges discussed at the end of the previous section, from an industry perspective there is reason to be optimistic about the near future for LNG. First, air-quality concerns have been a key driver of demand growth in recent years in some countries in Asia. This has meant that national governments have shown a willingness to use regulatory intervention to favour cleaner energy sources, including natural gas, in order to drastically reduce local air pollutants, such as NO_x, SO_x and particulates. With economic development, and the growth of a middle class that can afford to be more concerned about the environment, air quality starts to trump the costs of energy. It is likely that this phenomenon will occur in other emerging economies as well.

Second, and of critical importance, the entry barriers into the natural gas market are lower than they used to be. In cases where countries are depending on imports of natural gas, constructing major natural gas pipelines is capital-intensive, and geographically, environmentally and politically challenging, which explains why very few are either being built or under serious consideration. But, as the market for LNG continues to mature, other options are emerging. It is likely that the usage of FSRU technology will become more widespread. It is also important to note that, if countries cooperate, LNG can help to address regional challenges, not just national ones. To give one example, LNG imported through the FSRU in Lithuania has to date been consumed in four different countries (the three Baltic states, and Northern Poland). Once delivered, natural gas can flow to Latvia and Estonia by pipeline, and, absent pipeline infrastructure between Lithuania and Poland, transport has taken place by vessels and trucks. An even more recent trend in distribution of more modest volumes of LNG is ISO containers, which can be transported either by rail, sea or road. In 2018, ISO containers were used to supply Puerto Rico with LNG from nearby Florida, for example.[45] All of this points to greater flexibility of LNG trade at lower volumes, a welcome addition to the 'big is best' mentality that has long dominated LNG markets.

Third, new markets for natural gas are emerging, particularly in parts of the transportation sector. In China, urban air-quality regulations are forcing transport companies to invest in LNG trucks at the expense of more polluting diesel-powered trucks.[46] Finnish gas company Gasum Oy in 2017 announced it is expanding its LNG filling-station network in Scandinavia in order to incentivize long-distance trucking companies to switch fuels.[47] In the global shipping sector, stringent IMO SO_x emissions targets, which have come into effect in 2020, are forcing shipping companies to invest in scrubbers, or purchase low-sulphur oil products, or shift to cleaner-burning LNG. For a variety of reasons – including slow turnaround time of the shipping fleet and the time it takes for port cities to build up LNG bunkering facilities, and concerns about the life-cycle CO_2 emissions associated with LNG production and transportation (which are discussed in chapter 5) – it is anticipated that natural gas will only see a significant uptick in demand for shipping by the middle of the next decade and beyond.[48]

Fourth, there is some indication that major OECD countries

want to make good on their commitments under the Paris Climate Accord. As we discuss in more detail in the next chapter, it is uncertain what role natural gas will play in decarbonized economies in the long term. However, it does seem that, in the short and medium term, natural gas demand may receive a boost, as governments focus on reducing (particularly) CO_2 emissions by phasing out coal from power generation.

Fifth, there is reason to be moderately optimistic that industry can contain, and possibly further reduce, costs for natural gas in the future. Some analysts remain pessimistic about the costs of natural gas, in comparison to some of its competitors.[49] We would first reiterate that, in key growth markets, such as China, policy rather than price seems to be the key determinant of the growth rate of consumption. Second, we observe that companies continue to be able to reduce production costs.[50] In addition, considering the fierce competition amongst LNG project developers, we would speculate that this, in theory, should keep the lid on project economics as well.

Next to costs, the other critical uncertainty about the future of natural gas is its compatibility with an increasingly greenhouse-gas constrained world. Is natural gas a bridge fuel towards a low-carbon economy, with a clear ramp off at some point in the future? Or is it a destination fuel – for instance, because existing low-carbon technologies do not allow for a full substitution of GHG-intensive fuels, including natural gas, in the coming decades? We discuss this in the next and final chapters.

Conclusions

The LNG industry is undergoing a period of significant change, which Corbeau and Ledesma have called 'the great reconfiguration'.[51] After the shale gas boom that we described in the previous chapter, this is our second revolution. The traditional exclusive club of LNG buyers and sellers is no more, and the democratization of LNG is well under way. Several developments are contributing to and/or reinforcing this trend. First, there is abundant supply. Specifically, in the United States and Australia, the growth of natural gas production has, once (most) domestic demand had been met, incentivized companies to look for markets elsewhere. This has (amongst other things) spurred major investments in liquefaction capacity, and an increase in natural gas trade by ship. Seaborne trade will likely overtake international pipeline trade sometime in the not-too-distant future. Second, technology has helped to make LNG more accessible, and affordable. We described how FSRUs allow countries of relatively modest size to consider importing LNG, and how countries with credit challenges can now, sometimes with the help of development institutions, join the market. Consequently, the amount of countries that can import LNG has more than tripled in the last fifteen years. On the liquefaction side, technology is not set in stone either, and companies are trying to develop new business models, in an attempt to reduce costs (appreciating that the overwhelming majority of demand growth in the decades ahead will happen in emerging economies). In addition, FLNG projects can help bring resources in frontier areas – such as Mozambique, Mauritania/Senegal, Cameroon and Congo Brazzaville – to market. Third, concerns about local air quality, and policies to address those concerns, have been a major contributor to global demand growth. The most obvious example has been China, where government orders to replace coal in industrial and space-heating sectors with natural gas incentivized a surge in demand that started in 2017 and continues today, albeit at a reduced pace. Because energy demand growth will almost exclusively take place in industrializing non-OECD countries, and because hundreds of millions of people are projected to join the global middle classes, we anticipate that concerns about local air quality will continue to drive natural gas demand.

To be sure, there many uncertainties regarding the coming of age of LNG markets as well. Principally, it is to be determined how LNG will be priced in the future, and the fact that recent agreements have included references to oil products, coal, and natural gas trading hubs in various parts of the world is a nice illustration of this. Market liberalization in East Asia will be critically important in this regard, because a more competitive environment will allow for price formation based on local supply-and-demand dynamics. The jury is also out regarding whether long-term contracts between buyers and sellers will continue to be a feature of this market. Currently, some companies have financed new liquefaction projects off their balance sheets, but not all companies can do this, and without new financial models, those companies will continue to need firm long-term commitments from buyers to secure their loans. Will buyers continue to be interested in signing contracts, or will we reach a point where they feel that they can comfortably turn to the spot market without overpaying for LNG? If so, how will that impact the LNG market structure, number of participants, and competition? Time will tell. The number one long-term uncertainty though is related to GHG emissions. At the end of the day, natural gas is a fossil fuel: its production comes with fugitive methane, and its combustion releases a significant amount of carbon into the atmosphere.

For natural gas to secure its long-term role in global energy markets, a third revolution is necessary. This is the topic of the next chapter.

CHAPTER FIVE

The Future Role of Natural Gas

From a bird's-eye view, notwithstanding that it is a fossil fuel, the outlook for natural gas demand worldwide seems rosy. Various forecasts suggest growth rates in the short (2020s) and medium term (2030s) that other fossil fuels cannot match (see table 5.1). The reasons for this have been debated extensively, including in this volume. Critically, from an environmental perspective, when burnt natural gas releases roughly half the amount of carbon emissions that coal releases, and roughly a quarter of those of crude oil, this makes it a relatively clean fossil fuel. In addition, when produced and transported, ill-defined amounts of methane may slip into the atmosphere (fugitive emissions). These estimates are controversial, specifically because our understanding of these emissions throughout the production cycle is relatively modest, clouding our understanding of the exact GHG footprint of natural gas. Despite these uncertainties, there is little, if any, compelling evidence that the GHG footprint of natural gas is as bad as or worse than that of its peers, and initial action is now being taken to reduce life-cycle emissions, which, if successful, can help make it a viable fuel for the longer term, as the world transitions to a low-carbon energy system.[1]

As we have noted throughout this volume, natural gas will continue to play different roles in different parts of the global energy economy for some time to come, both as a source of power and heat, as a fuel in transport, and as a raw material in industry, and different narratives will continue to coexist in various parts of the world. Too often the role of natural gas in the decades ahead is portrayed as a zero-sum issue. In these public and political debates, natural gas is either the best thing that has happened since sliced bread, or a Faustian bargain that will bring us doom and gloom as investments today lock us into a level of future carbon intensity that does not match our assessments of the speed at which decarbonization must take place in the decades ahead in order to stem the worst impacts of global warming. Increasingly, we are warned that new investments in exploration and production are not compatible with the objectives of climate and environmental policy, and any such investments could end up stranded in the foreseeable future.[2] Reality is substantially more nuanced, and this is the central topic that we explore in this chapter. We argue that it is too early to signal the long-term demise of natural gas, but observe that a third revolution is required to safeguard its possible role as a fuel of the future.

The key drivers of future gas demand

It is important to remember that, in addition to the benefits of natural gas in terms of GHG emissions, there are clear and undisputed benefits in terms of local air pollutants. Usage of natural gas comes with modest amounts of NO_x, negligible amounts of SO_x, and particulate matter – all key contributors to smog and poor air quality. The World Health Organization estimates that, annually, 4.2 million premature deaths are attributable to outdoor air pollution. Another 3.8 million people die prematurely as a result of household exposure to smoke from dirty cookstoves and fuels.[3] Thus, there are two key drivers (co-benefits) of investments in natural gas.

The first hinges on the idea that decarbonization takes time

Table 5.1 Global gas demand scenarios (bcm)

Source	Published	2020	2025	2030	2035	2040	2050
IEA WEO 2018 (New Policies Scenario)	Nov. 2018	3635	4293	4641	5025	5399	–
IEA WEO 2018 (Current Policies)	Nov. 2018	3635	4386	4860	5366	5847	–
IEA WEO 2018 (Sustainable Development Scenario)	Nov. 2018	3635	4189	4318	4298	4184	–
BP Outlook 2019 – Evolving Transition Scenario*	Feb. 2019	3927	4345	4609	–	5229	–
BP Outlook 2019 – Rapid Transition Scenario*	Feb. 2019	–	4486	4088	–	4056	–
ExxonMobil Outlook for Energy 2018**	Feb. 2018	3824	4164	4473	4705	4908	–
Equinor Energy Perspectives 2019 Renewal Scenario	June 2019	–	–	4246	–	–	3182
Equinor Energy Perspectives 2019 Reform Scenario	June 2019	–	–	4471	–	–	4787
Equinor Energy Perspectives 2019 Rivalry Scenario	June 2019	–	–	4422	–	–	4765
DNV GL Energy Transition Outlook 2018	Oct. 2018	4116	4346	4742	4835	4650	–
EIA Reference Scenario 2018 ***	July 2018	–	–	–	–	5096	–
Shell Sky Scenario****	Mar. 2018	3967	3942	4115	4122	3920	2991

Note: according to BP, actual consumption in 2018 was 3848.9 bcm.

* Mtoe converted to bcm using a conversion factor of 1.11

** QBtu converted to bcm using a conversion factor of 28

*** tcf converted to bcm using a conversion factor of 28.32

**** EJ converted to bcm using a conversion factor rate of 26

Source: updated from BEIS (2020), *Fossil Fuel Price Assumption 2019*. London: BEIS.

– how much time is a matter of debate and disagreement,[4] but for the moment intermittent renewable energy sources such as solar and wind will continue to need firm and flexible back-up power in order to balance the system. In addition, to date – in most instances – the impressive growth of wind and solar energy have not come at the expense of fossil fuels. Rather, they have supplemented existing energy demand, illustrating continued energy demand growth and the rise of global middle classes.[5] Proponents of natural gas will then argue that the fuel is not perfect, and surely not a climate policy per se, yet it can be amongst the best available options on a path (often described as a bridge) to a low-carbon economy, in terms of cost-effectiveness and public acceptability, but only in certain contexts. For example, the IEA estimates that, worldwide, a proposed 1,200 metric tons of CO_2 could be abated by switching from coal to natural gas in the electricity sector, the vast majority of this to be realized in the EU and US, where installed generation capacity is relatively old.[6] Furthermore, if carbon capture, utilization and storage (CCUS) technologies were commercialized and scaled, both gas-fired power stations and industrial plants could significantly reduce their carbon intensity. As we shall see, the availability of CCUS also makes possible the use of natural gas as a feedstock for a hydrogen energy system. Sceptics dispute this argument because significant uncertainties remain about the commercialization of CCUS technologies, which to date have not lived up to expectations. Nonetheless, these technologies continue to be baked into most climate modelling work, particularly in combination with biomass as a source of negative emissions, and are broadly acknowledged as being a necessary tool in a wider mitigation portfolio to achieve large-scale carbon emissions reductions, in a cost-effective manner.[7] Ironically, given the limited interest of the Trump Administration in climate change, in 2019, of the eighteen CCUS projects worldwide, eight are in the United States.[8] Sceptics would also point to mounting concerns about fugitive methane emissions. Clearly, these emissions have been underappreciated to date, specifically in the upstream sector (at the wellhead, and gathering lines). A recent study suggests that US EPA estimates of fugitive methane from US oil and gas systems may be 60% lower than actual emissions rates.[9] Even then, at such a level, natural gas still delivers a decarbonization benefit in comparison to coal (the literature broadly agrees the threshold of leaked methane to be around 3.2%, or possibly a little higher depending on your timescale[10]), but these data pose a significant challenge for the natural gas industry. Ironically, but understandably given the enormous increase in the number of new wells being drilled, it is the shale gas revolution that has heightened our awareness of the challenge of fugitive methane emissions. The IEA estimates that, in 2017, total oil- and gas-related emissions were 80 million tons, and that more than half of that came from natural gas operations, notwithstanding the high level of uncertainty around such emissions, largely due to a lack of good data.[11] Furthermore, the IEA estimates that, of fugitive emissions, an approximate 45% can be captured at net zero cost, using existing technologies such as vapour recovery units and leak detection and repair programmes. However, by 2030, a 75% reduction in fugitive methane emissions is necessary in order to meet the targets under the IEA Sustainable Development Scenario, and currently industry is not on track to make those reductions. Fugitive methane emissions also form a challenge for the LNG industry, which inevitably has a significantly higher emissions intensity than pipeline gas due to the need to use large amounts of energy to compress and liquefy the gas, to transport it and then to vaporize it once unloaded. A recent industry study suggested that 'the losses (largely as fuel) of natural gas along the

LNG chain can account for more than 12–13% of the original hydrocarbon gas produced at the wellhead; this is compared with less than 1% for a typical pipeline project'.[12] However, it is interesting to note that the IEA maintains that 'LNG imports to China result in fewer emissions than pipeline imports, as a consequence of fugitive methane emissions that occur along the value chains'.[13] The situation is even more complex as some supply chains now combine lengthy pipeline transmission with liquefaction. This is the case in the US where the new LNG terminals draw on the national gas delivery system rather than a dedicated field. This makes the calculation of carbon intensity and life-cycle emissions very complex.[14] Moreover, it is worth keeping in mind that, at the end-user stage, we must assess what alternative fuel would have been used if it were not natural gas. The precise GHG footprint of shipping a cargo of LNG to China may be poorly understood, but that does not mean that using that natural gas to replace coal in industrial heating, for example, does not come with societal benefits, specifically measured in terms of a substantial reduction in local air pollution, resulting in measurable health benefits. The bottom line is that, while the life-cycle emissions of the natural gas supply chain may not negate its credentials as a relatively clean fossil fuel, they have certainly tarnished its reputation and there is an urgent need both to accurately account for those emissions and to improve regulatory and industrial practices to reduce them as much as possible.

Zooming into the United States, with over 100,000 producing wells and tens of thousands of kilometres of pipeline infrastructure, collecting good data is easier said than done, especially when it is not a regulatory requirement. However, in light of recent estimates, the scale of fugitive methane emissions is such that, within industry, a sentiment is growing that it cannot afford to ignore the issue while maintaining that natural gas has a role to play in the transition towards a low-carbon economy. There are now various industry initiatives in place to reduce fugitive emissions, and individual project developers are increasingly concerned about the life-cycle emissions associated with their supply chain.[15] Still, the IEA suggests that leakage levels are likely higher in countries where regulation is lax, and observes that in countries such as Canada, and multiple US states and EU countries, mandatory targets to reduce fugitive methane from oil and gas systems are implemented. Going forward, not only better measurement, but also data verification and transparent reporting will be critically important.

The second driver of natural gas demand centres around the problem of local air pollution, and in Asia this is a far more significant driver of gas demand growth than decarbonization. Natural gas can play a role in backing out more polluting fuels – be it coal, oil products such as diesel and fuel oil, or traditional biomass – and improving local air quality, thus increasing life expectancy. This is not just a matter of substitution in power generation. Rather, natural gas is an increasingly popular fuel for industrial use and space-heating, as we have seen in the case of China. In addition, in countries where traditional biomass is mostly used for cooking, switching to gas cylinders can lead to a major health improvement. To be sure, these cylinders will mostly contain propane or butane (or a combination hereof), also known as LPG, and mostly not CNG. The reason is simple: since the energy content of NGLs is higher, cylinders with propane or butane will last longer. CNG has proven to be a good alternative to oil products in the transport sector – as has, for instance, been demonstrated in major urban areas in India.[16] Finally, LNG offers an alternative to the most polluting oil products in other parts of the transportation sector, such as long-distance trucking, and shipping.

As noted above, concerns about clean air prevail in emerging economies, where coal, fuel oil and traditional biomass are a major threat to air quality and human health. This is something that is often lost in the public discourse in the OECD world. The fact that the larger part of the world is at a different stage of development impacts energy transition discourses as well. Policy debates in parts of the EU and the United States increasingly centre around the notion of moving beyond natural gas. Yet it is unlikely that one will pick up this narrative in Beijing, Mumbai, Jakarta and the like, where the vast majority of future energy demand growth lies, and where the contribution that natural gas can make to improving local air quality is measurable and positive overall, and discussions about deep decarbonization are rarely on top of the policy agenda.

Importantly, also in parts of the 'developed world' where polluting fuels such as coal still prevail, this driver of local air-quality improvement may be more likely to gain traction than the one around GHG emissions reductions. One example is Poland, where thirty-six of the fifty most polluted cities in the EU are located,[17] and where the national government has been deeply sceptical of progressive EU policy proposals to curtail GHG emissions (or of replacing coal with natural gas). Even in countries where gas has supplanted coal – such as the UK – the environmental benefits it has brought are not always acknowledged. Similar benefits are now measurable in the US as natural gas continues to force coal out of the power-generation mix: between 1997 and 2014, emissions of SO_2 and NO_x were reduced by 44% and 40%, respectively.[18] Of course, the effects of fuel switching on carbon emissions reductions have been reported extensively. Between 2005 and 2017, US carbon emissions from the power sector fell by 28% due to slower demand growth for electricity (in part because of a major economic recession) and changes in the fuel mix. Those changes occurred due to coal-to-gas switching and the growth of renewable power – e.g. solar-PV and onshore wind. In 2016, natural gas surpassed coal as the largest source of power generation.[19] Specifically, natural gas accounted for one-third (35%) of utility-scale electricity generation in 2018, followed by coal (27%), nuclear (19%) and hydropower (7%).[20] In sum, while there seems to be a legitimate value to natural gas as a cleaner burning fuel (which seems important in an era where backing out the most polluting fuels is increasingly a top-tier policy concern), as we discuss below, the jury is still out on the role that it will play in a future low-carbon economy in various parts of the world.[21]

A transition or destination fuel?

Absent widespread commercial deployment of technologies such as CCUS to mitigate the negative externalities that come with burning natural gas, it is evident that, as the world transitions to a low-carbon economy, there is an end-date for the growth of natural gas (most modelling suggests some time in the 2030s at a global scale, with some regions not seeing a decline until the 2040s). Those making the case for natural gas as a 'destination fuel' implicitly assume either that technologies such as CCUS will be commercially available, or that the GHG emissions reduction targets, as agreed under the 2015 Paris Agreement, will not be met. It is worth noting that current national commitments under that accord are not yet compatible with a 'less than 2 degrees' world.[22] Furthermore, as noted above, many of the scenarios that are aligned with the Paris Agreement make implicit assumptions about the availability of negative emissions technologies, particularly biomass and CCS (BECCS), despite there being significant uncertainty about whether those technologies will in fact be commercially available, and at scale. Given these uncertainties, investments in natural gas production and its associated infrastructure arguably come with a 'transition risk' factor that will only grow in the years and decades ahead. This is important because these investments are long-term and capital-intensive, with projects having to pay off over multiple decades. Thus, the lifetime of projects that companies invest in now may well extend beyond the moment where, according to climate models, natural gas usage should start to decline. In its Sustainable Development Scenario, the IEA foresees that natural gas demand in the United States and the EU by 2040 will have declined by an estimated 120 bcm and 170 bcm respectively, while demand growth in China will have levelled off by then, as demand in India continues to grow. Other research suggested more modest demand decline rates. The INGAA Foundation, in its Rapid Renewables Transition Scenario, finds modest declines in natural gas demand in the United States by 2040 in both residential and commercial sectors, and industry, with 0.3 per cent and 0.2 per cent per year, respectively.[23] In this scenario, global LNG demand declines as renewables and nuclear energy make inroads in Asia, curtailing US exports of LNG around 10 bcf/d. Pipeline exports to Mexico stabilize around 6 bcf/d.

Looking at various regions in the world, substantial differences in the current status and prospects for natural gas can be identified. Research by the UKERC that examined the long-term potential role of natural gas reached the following conclusions. First, that a period of increased gas consumption must occur alongside a much greater reduction in coal consumption. Second, the period of increased gas consumption is strictly time-limited. Third, methane leakage throughout the natural gas supply chain must be limited. Fourth, gas use must increasingly be coupled with CCUS (in the absence of CCUS, gas consumption must peak and decline much more rapidly, or methane must be replaced with alternatives like biomethane). Finally, there is significant sectoral and regional variation in the potential for natural gas to aid in the transition to a low-carbon energy system, with its main sectoral role being to aid decarbonization in power generation and industry (although, as noted above, air pollution is currently the key driver of the surge in gas demand in Asia), and in geographic regions that are currently reliant on coal.[24]

With the rise of shale gas, natural gas in the United States has grown very rapidly over the course of the last decade, replacing coal as the primary fuel for power generation, triggering major industrial investments,[25] and incentivizing a massive expansion of both pipeline and LNG export capacity. Here, abundant and cheap gas means that there are no energy security concerns associated with increased use of natural gas. However, there are growing concerns in relation to the cumulative impact of shale gas production on human health and the natural environment, as reflected in the increasingly critical narrative about natural gas as part of the political spectrum. In 2013, the then Mayor of New York, Bloomberg, made a compelling case in favour of natural gas in a speech at Columbia University.[26] However, in June 2019, that same mayor announced a campaign to curtail further the use of coal in the United States, and halt the growth in natural gas production, and consumption.[27] Thus, one can argue that a 'bridge' to a more sustainable future has been built in North America, yet queries about where the off-ramp might be grow increasingly vocal. President Obama and Energy Secretary Moniz made attempts to address the key environmental concerns linked to the build-out of natural gas, specifically by designing regulations aimed to curtail fugitive methane emissions and promote the development of CCUS technology. In turn, President Trump is attempting to roll back the aforementioned methane regulations. However, in a rare moment of bipartisanship in Washington, DC, Republican and Democratic lawmakers continue to push for incentives for companies to invest in CCUS technology.[28] Certain environmental groups argue that, in some contexts, the dash for gas has constrained the growth of renewable energy.[29] Others point to the fact that it has sounded the death knell for the rejuvenation of the nuclear power industry. Indeed, there are several examples in both the United States and the EU where existing nuclear capacity has been, or is being planned to be, shut down, and replaced with natural gas, with evident negative consequences in terms of GHG emissions.[30] Thus, should the need for decarbonization demand a move away from natural gas, the US energy system could find itself ill prepared to deliver deep decarbonization without destroying large amounts of capital.

In Latin America, natural gas has a sizeable – and possibly larger – role to play in the future, although much will depend on the status of existing hydropower and the pace of renewable power development. In countries such as Brazil, natural gas has mainly been a supplement in power generation, depending on the availability of hydropower. In others, such as Argentina and Bolivia, natural gas has historically played a more prominent role, but the petroleum sector has been plagued by mismanagement and corruption. In recent years, investments in upstream activities across the continent, from Brazil and Argentina to Colombia and Peru, have picked up, with initial success most notable in Argentina, where imports of natural gas in the form of LNG have been turned into exports as companies have started to tap the massive Vaca Muerta shale formation.[31]

The EU, though being one of the globe's main markets for natural gas, has developed a complicated relationship with this fuel, chiefly because of environmental and geopolitical issues (explained in chapters 2 and 3). In several member states, concerns about the timely achievement of carbon emissions reductions under the Paris Agreement are putting natural gas in a squeeze, as the confidence that the negative externalities that come with using natural gas can be mitigated has waned.[32] The Netherlands stands out because, in addition to the aforementioned sentiment, domestic production has been shut in following a series of earthquakes that were linked to natural gas production in the country.[33] In many EU member states, incentivizing new exploration, in particular the development of shale gas (as discussed in chapter 4), has been complicated by growing environmental concerns. In other member states, environmental concerns may feature less prominently, but geopolitical considerations are important. Poland is the best example of a country that has been unwilling to endorse an increased reliance on natural gas at the expense of more polluting coal.[34] Instead, it has developed a strategy that leans heavily on infrastructure that needs state money and financial support from the EU to get built, in order to replace Russian natural gas in the country with resources from elsewhere, such as LNG and pipeline gas from Norway. The country's political leadership also continues to flirt with the possibility of developing nuclear power, even though it is unclear how these investments would be financed, and who would construct these plants. In Southern Europe, debates about natural gas are less entrenched with geopolitics, though this may change as the leadership struggle in Algeria, a main supplier of natural gas to the region, unfolds, and the war in Libya continues. Then there is the promise of substantial additional production in the East Mediterranean, but the local politics are complicated. The recent development of natural gas fields offshore of Israel and Egypt has already been transformational for these countries, but only time will tell whether the broader region can benefit in full. On the upside, European companies and member states have invested heavily in infrastructure to be able to import various sources of natural gas supply, which seems important now that domestic production is declining rapidly, and several countries are slowly starting to make good on their promises to reduce their use of the most polluting fuels to improve local air quality, and their promises under the Paris Climate Accord. In 2019, close to ten EU member states have announced 'end dates' for coal – e.g. France by 2021, the United Kingdom by 2025, and several EU member states by 2030. Germany, the continent's largest coal consumer, is attempting to formalize its rather modest ambition to end coal use no later than 2038. However, once these announcements have been made, investors may look for a gracious way out, and coal may be phased out sooner than planned.[35]

Decarbonizing natural gas

With geopolitical concerns still looming large, and with growing public impatience with the pace of the energy transition, it seems unlikely that it will be smooth sailing for natural gas in the EU going forward. The case of the United Kingdom, historically a front runner in the EU on both climate and energy policy, is illustrative. The British were the first to implement unbundling rules in both the natural gas and electricity sectors, and were amongst the most progressive (and, in fact, successful) EU member states when it comes to carbon emissions reduction, as illustrated by their unilateral implementation of a carbon price floor under the Emission Trading System (ETS), which comes on top of the ETS price and the recent approval of a net zero emissions target for 2050.[36] Thus, effectively, the UK has a carbon price of around $50 (though this could be reduced post-Brexit), which in turn has wiped much of the existing coal-fired power-generation capacity from the merit order, substantially contributing to carbon emissions reductions in recent years. The government has now determined that by 2025 there will be no unabated coal-fired power generation in the UK, but it may well be gone earlier due to prevailing market and regulatory forces. However, those who believed that such policies were good news for natural gas have been disappointed (this included the UK government when it announced the phase-out of coal) as demand continues to fall, in part due to improvements in the efficiency of domestic gas boilers, but also due to competition from renewable power generation. According to BP statistics, in 2004 UK natural gas consumption peaked at 94.7 bcm, but by 2018 it had fallen by 16.8 per cent to 78.8 bcm. There is ongoing debate about the role that natural gas may or may not play in the long term in the UK and elsewhere in Europe, and most of this discussion centres around the potential for hydrogen, and biomethane.

There are those in the UK – principally the owners of the gas networks – who believe that natural gas can be used as a feedstock to produce hydrogen through steam methane reforming (SMR), which in turn could replace natural gas for heating and cooking, and in transport. The city of Leeds is, in collaboration with industry and local government, developing an initiative to this effect, which explores both the technical and commercial feasibility of such an effort. Thus, if this project were to be implemented in full, effectively, after four decades, Leeds would return to the old days of 'town gas'.[37] This time the city gas plant would be producing hydrogen, and the associated CO_2 transported to the coast by pipeline to be sequestered in salt caverns in the North Sea. Consequently, CCUS is receiving renewed attention in the UK's Clean Growth Strategy, after being delivered a serious setback in the UK in 2015, when the government at the eleventh hour decided to pull financial support for a research initiative that was designed to help commercialize CCUS technology at scale in the UK.[38]

Alternatively, most analyses assume a massive build-out of renewable power in the years and decades ahead, supported by an increasingly smart grid and electricity storage technology. However, there are uncertainties about the limitations of existing electricity storage technologies, and the variability of renewable energy. Consequently, many scientists focus on using excess renewable power to produce (green) hydrogen through electrolysis. This is very much the Continental European view of the future role for hydrogen. Essentially, in this process, renewable power is used to split water into hydrogen (H_2) and oxygen (O_2). The H_2 is then converged with CO or CO_2 to produce methane (CH_4), a process called methanation.[39] One advantage is that this does not result in CO_2 emissions (and thus does not require CCUS to be carbon neutral).[40] Another advantage is that the gas grid can help absorb excess renewable electricity when necessary, and, provided that the hydrogen can be stored, provide a back-up when electricity demand is higher than supply. Subsequently, these power-to-gas technologies could also help minimize grid expansion as the amount of renewable electricity grows. However, there are also major challenges, including issues mixing H_2 and CH_4 and maintaining system stability (consequently, H_2 injection in the grid is currently belived to be limited at anywhere up to 12 per cent),[41] and, most importantly, high costs, making power-to-gas plants unprofitable under current market conditions.[42] Therefore, some have concluded that power-to-gas technologies are primarily an option to deal with excess renewable electricity, rather than replacing current demand for natural gas.[43]

The 'gas pool' could be further expanded by promoting the development of biomethane (biogas processed to pipeline specification) and biosyngas (which uses biomass or prepared waste to produce synthetic methane via a thermogenic process). Agricultural and municipal waste or other organic material can be converted into biogas in a process called anaerobic digestion. This biogas can be used to produce electricity, heat (often as combined heat and power / CHP plants), or in transport. To upgrade biogas to biomethane, contaminants like hydrogen sulphide (H_2S) and ammonia (NH_3) have to be removed. To increase the energy content, CO_2 is removed until the same Wobbe index is reached as natural gas. These technologies will have to be commercially proven in the decades ahead and may be limited in scale due to the availability of feedstock, particularly if the focus is exclusively on waste and residue.[44] Still, there is considerable potential: the EU had over 17,000 biogas plants in 2015, representing more than 10 GW of electricity-generation capacity (globally the total was 16 GW at the time). However, to date, the growth in biogas is based mostly on subsidized electricity generation, and the use of heat is generally limited.[45] Here, too, the key question is: can the economics of these projects be improved? In India, most of the potential for biogas (estimated at between 29 and 48 bcm per annum) has not been harvested, due to high capital costs, credit challenges of municipal organizations, poor waste management and lack of a steady feedstock.[46] At this stage, it is not clear how industrial heat would be provided, or how energy-intensive industries such as cement- and steel-making would continue to operate. There are many promising pilots under way for low-emissions steel and cement production, but all are at an early stage of development.[47] As long as the jury is out on technologies like these, decision-makers in both the public and private sectors will continue to seek a balance between emissions reductions and responding to short-term pressures (e.g. from shareholders, politicians or the broader public).

In these scenarios, will natural gas become merely a back-up fuel for periods when renewable energy cannot meet all demand? Some have suggested that, in the absence of CCUS, natural gas demand in the UK in 2050 could fall to 10 per cent of 2015 levels.[48] Interestingly, the UK Climate Change Committee's recent Net-Zero analysis, which proposes the development of hydrogen and a very modest role for gas in power generation, concludes that gas demand in 2050 could be at two-thirds of the present level.[49] Table 5.1 illustrates the differences in the outlook for natural gas demand, depending on how swiftly the energy transition progresses. Scenario studies like the one by UKERC, which concluded that there may only be a modest role for natural gas in the medium to long term, raise new questions – for instance, about commercial models for sustaining adequate levels of natural gas production and investment in infrastructure, in the context of heavily fluctuating demand throughout the year. Who would invest in gas-fired generation capacity that only serves peak load? Thus, in liberalized markets like parts of the EU and the United States, these developments may also impact on the division of labour between private and public sectors as we know it today.[50] Put simply, can the market deliver deep decarbonization without significant state regulation, and even ownership? Clearly, next to major technological challenges, there are uncertainties about scaling-up technologies, making them commercial, and finding ways to share project risk in a world with significant uncertainties.

This discussion about the 'decarbonization' of natural gas and the future of various different 'gases' in 2019 may feel profoundly Northwest European, even though there are hydrogen initiatives under way in other parts of the world, such as Japan (which championed it during their presidency of the G20 in 2019) and South Korea, as well. Elsewhere – for example, in

Canada and the United States – these discussions typically seem to be more niche now, due in no small part to the availability of abundant low-cost natural gas, and, in the case of the US, a federal government with other priorities. However, in US states with progressive climate policies, discussions about the fuel mix 'beyond natural gas' have emerged in places like New York and California, and biomethane or hydrogen projects may well become part of states' Renewable Portfolio Standards in the not-too-distant future. In addition, state regulatory authorities are, in light of continuously falling costs of renewable and electricity storage technologies, increasingly reluctant to support new investments in fossil fuels, including natural gas.[51] In the European context, it is important to note that a group of transmission system operators (TSOs) – such as National Grid, Snam, Gasunie and Fluxys – are playing a prominent and proactive role in discussions about 'greening' the natural gas mix in the decades ahead. In essence, these TSOs intend to prepare for a switch from methane to cleaner options (biomethane and biosyngas) and green hydrogen at some point in the future. Although this has not been officially stated, it is reasonable to assume that this focus is in part incentivized by scattered public calls (often at the local governance level) to dismantle distribution grids for natural gas altogether, in order to curtail usage of the fuel. The city of Amsterdam in the Netherlands is one example where local leaders have vowed to phase out natural gas altogether, though it is important to place this in the context of the complicated relationship that the Netherlands has recently developed with natural gas.[52] Thus, the current focus of TSOs on alternative gases such as hydrogen and biogas can be seen as a genuine attempt to contribute to energy system decarbonization, and a survival strategy to preserve the value of their networks and a social licence to operate.

Despite receiving growing attention, concerns about the

long-term compatibility of natural gas with the EU's desire to reduce GHG emissions, including a desire to commit politically to net-zero emissions by 2050, are often trumped by the perpetual short-term concerns about energy security and geopolitics. Thus, it is too early to tell whether the European efforts to grow very significantly the share of hydrogen and biomethane are going to be successful. It is also uncertain to what extent these efforts will take root in other parts of the world – though, over time, they likely will. If successful, these efforts will markedly affect current thinking about future demand for natural gas, and thus further cloud the investment climate in the years ahead. This is reinforced by growing pressure from shareholders and financial institutions to bring current and future investments in line with pathways to deep decarbonization.

Conclusions

On balance, potentially, natural gas has more of a future in the non-OECD world, but there are several conditionalities, which at this point in time are of varying significance, depending where the assessment is made. These conditionalities centre around the following questions: first, can the natural gas industry substantially curtail fugitive methane emissions? Will the larger part of the industry endorse progressive rule-making on this matter, setting ambitious targets while leaving room for technological innovation? Or will this issue prove divisive, resulting, after a period of muddling through, in more stringent rules that might harm business operations? Second, can CCUS technologies be commercialized, and developed at scale? The promise of these technologies, both as an effective mitigation tool, and as a cost-effective way to reduce CO_2 emissions, has long been acknowledged. Will sufficient countries and companies show leadership in the years ahead to develop these technologies at scale? Third, can 'greener gases' make a significant contribution to the fuel mix in the decades ahead? Can the costs of various projects be reduced to make hydrogen cost-competitive; and will companies be able to commercialize CCUS to enable net-zero blue hydrogen production? Will power-to-gas technologies be able to do more than function as a buffer for intermittent renewable electricity? Can large quantities of biomass be turned into biomethane, with technical hurdles addressed and costs reduced? In short, after the shale revolution and the coming of age of LNG, a third revolution is needed to develop 'natural gas 2.0' in the 2020s to preserve a role for natural gas in the 2030s and beyond.

Here, it is important to note that currently the conditionalities outlined above apply exclusively to OECD countries. As we have discussed throughout this volume, natural gas demand growth, and the vast majority of energy demand growth more generally, is anticipated to take place in the non-OECD countries. Unsurprisingly, different conditionalities apply for natural gas to be successful in that different context. Key questions may include: will governments continue, and possibly intensify, their policies to address local air pollution, as more people join the middle classes, and can afford to care about cleaner air? Can new technologies be developed (or transferred), and commercialized, to help distribute natural gas more widely, in parallel with major investments in cleaner energy technologies? Can companies (if need be, with state support) control and preferably reduce the costs of delivery of natural gas, to cater for the hundreds of millions of people who have no access to electricity, and the additional hundreds of millions that have access only to traditional biomass?

Considering all these uncertainties, it is a challenge to predict future demand for natural gas, and subsequently to attract the necessary capital, to make profitable investments.

Logically, companies will increasingly focus on developing the lowest-cost resources. One reason for the growing difficulty of portraying a unified picture of 'natural gas 2.0' is that it will likely differ widely in various geographies around the world. Decarbonization is increasingly pursued in parallel with decentralization in an age where national-level leadership struggles to get successful policies in place, which raises questions about the division of the costs and benefits of the energy transition, as well as social equity. In this landscape, for the short to medium term at least – into the 2030s – natural gas will continue to make sense in most parts of the world. For those parts of the world where natural gas is chosen as a fuel of choice for the decades ahead, questions about mitigation strategies and technology will be voiced more often going forward, by environmentalists, the broader public, shareholders and the financial community. For natural gas to have a long-lasting future, an approach of letting a thousand flowers bloom seems logical. Smaller-scale markets, more trade, more liquidity, cost reductions and new business models with an overall focus on drastically curtailing life-cycle GHG emissions must all be pursued to find a future role for gas in a decarbonizing world.

CHAPTER SIX

The Golden Age of Gas?

In classical Greek mythology, a Golden Age was the first and best period of a world, a period of prosperity, generally preceded by the decline of another world. Historians argue about how a Golden Age comes to an end, but one possible explanation offered by the Greeks was *hubris*, or the overstepping of bounds. In 2011, the IEA asked what the world would look like if there were a 'Golden Age of Gas'.[1] They developed an alternative scenario, based on their New Policies Scenario at that time, which explored a future where gas played a greater role in the energy mix. By this, they meant that, between 2010 and 2035, gas consumption would rise by more than 50 per cent to meet 25 per cent of world energy demand in 2035. This suggestion was greeted with initial scepticism, with doubts mostly centring around the competitiveness of the fuel in comparison to its main peers, and questions about its reconciliation with the desired low-GHG energy systems of the future.

Speaking at the LNG Producer–Consumer Conference in Tokyo in September 2019, Fatih Birol, Executive Director of the IEA, observed that: 'The IEA's view that natural gas might be poised to enter a ' "Golden Age" was met with scepticism in some places, but many of the trends identified in 2011 have come to fruition.'[2] If we compare the IEA forecast of global gas demand in 2025 in their Golden Age Scenario (GAS) with the New Policies Scenario (NPS) in their *World Energy Outlook*

2018, the GAS scenario saw global gas demand at 4,384 bcm in 2025, while the NPS sees demand at 4,293 bcm in 2025. This suggests that the GAS scenario has actually become mainstream. The accelerated growth of demand in China has been critical. The GAS forecast that China's demand would be 247 bcm in 2015, reaching 335 bcm in 2020, and 430 bcm by 2025. According to BP's data, the actual level of gas demand in 2015 was 194.7 bcm, well short of the GAS forecast, but by 2018 it had surged to 283 bcm. The IEA's latest five-year gas forecast suggests that China's demand will reach 334 bcm by 2020 and 450 bcm by 2024 (well above the 2025 GAS forecast).[3] If 2018 demand grows at 10 per cent a year in 2019–20, it will reach 342 bcm. The US has also played a major role, and its demand growth is significantly ahead of the GAS forecast. The 2015 GAS forecast for the US was 661 bcm, but BP's data show that actual demand was 743.6 bcm. The 2020 GAS forecast was 668 bcm, but according to BP the actual level of demand in 2018 was already 817.1 bcm. The IEA's five-year forecast sees US demand at 895 bcm in 2024. There is one place where the Golden Age has struggled to materialize and that is Europe. Here, recent and forecast demand levels are well below those of the GAS forecast. This scenario saw EU demand at 553 bcm in 2015, and BP data show that it was 418.7 bcm. The GAS 2020 forecast was 587 bcm, but the IEA's five-year forecast sees it at 473 bcm. This highlights the importance of regional differences in the global gas story.

In this volume, we have argued that this current Golden Age has only been possible due to two 'revolutions'. Furthermore, for natural gas to preserve its long-term role in the global energy system, we argue that a third revolution is necessary.

The first revolution has been the rise of shale gas, CBM and tight gas, often lumped together as 'unconventional gas'. The large-scale application of existing production technologies to extract methane from rock layers has fundamentally challenged our understanding of the quantity of commercially available natural gas. Continued conventional discoveries in places as diverse as South Africa, Egypt, Indonesia and the Shetland Islands have further reinforced the belief that natural gas can be commercially produced for many decades to come. Consequently, buyers can credibly turn to natural gas as a resource – albeit for power generation, manufacturing, heating or transport – without being concerned about its physical availability. In addition, the rise of shale gas in the United States has allowed the world's largest economy to make a substantial dent in its emissions of carbon and local air pollutants, notwithstanding the political flavour of the day. It is evident that, absent drastic market intervention, coal will not make a comeback in the United States, simply because it cannot compete on price. Instead, natural gas is now so abundant that massive investments have been made to export the resource from the United States, either by pipeline (mostly to Mexico) or, increasingly, in the form of LNG.

The second revolution has been the coming of age of LNG. For decades, LNG was a rather exclusive market, mostly because cooling natural gas and building specialized ships to transport it are relatively expensive. However, in response to the Oil Crises, and further spurred by clean air policies, companies from Japan, the Republic of Korea and later Taiwan became large-scale importers of LNG, as they lacked domestic resources and opportunities to import natural gas by pipeline. Several countries in what would later become the EU also started importing LNG, and more have followed since, and there are now forty-three LNG-importing countries. Yet the real revolution for LNG markets has been the successful design and application of technology that is smaller in scale than that which the industry had grown accustomed to since the 1990s. Since 2004, FSRUs have created an opportunity for emerging economies to join the market for LNG as well. FSRUs are relatively modest in size, making them attractive for catering to markets that are smaller – but, most importantly, they are therefore easier to finance, if necessary, with the help of development banks. In recent years, this technology has proved successful in unlocking new demand for natural gas, despite a variety of economic and/or governance challenges in many emerging economies. At the same time, FLNG is becoming critically important to help monetize resources in frontier countries, such as Mozambique, Mauritania/Senegal, Cameroon and Congo-Brazzaville. With their continued rise, LNG suppliers are also challenging the status quo in mature markets, specifically in the EU and East Asia. In the former, incumbent pipeline suppliers such as Gazprom have been forced to adapt to new market realities, in an ever more competitive market environment, supported by strong regulatory oversight. In East Asia, several attempts to reform markets are under way in pursuit of more competition and lower prices, incentivized by developments in North America. A growing, and increasingly diverse, group of suppliers is vying for the business of buyers in these countries.

Though energy demand growth in mature markets is typically relatively modest, a key question in the years ahead is whether countries will make good on their pledges under the Paris Climate Accord and start curtailing their use of the most polluting fuels, possibly creating opportunities for natural gas. The United States stands out because, there, the rise of shale gas has effectively made the resource a transition fuel, with demand growth figures that are unparalleled for a developed economy. However, natural gas is not uncontroversial, and calls to halt its growth, or even move away from it, are increasing. It is noteworthy that a number of the Democratic candidates for the 2020 US Presidential elections have vowed to ban hydraulic fracturing. This is because of concerns about climate change, but also the ongoing build-out of natural gas infrastructure, and a major increase in flaring. We have reason to believe that in other developed economies (the EU, OECD Asia), natural gas is gaining a little ground and there is opportunity for fuel switching in the power sector; however, securing a long-term role in the energy mix requires a third revolution. In the short term, in addition to fuel switching, major reductions in fugitive methane emissions throughout the production cycle will be necessary for natural gas producers to be able to claim credibly that their product has a role to play in energy economies where GHG emissions are increasingly curtailed. In the medium and long term, large-scale commercial application CCUS technologies in power generation and industry will be necessary. Biomethane and power-to-gas technologies hold promise to become substantial additions to the gas mix, but technical, governance and, in particular, economic uncertainties, will have to be ironed out in the years ahead.

Even if this third revolution takes place, a major role for natural gas will still not be guaranteed. Competing technologies are gaining ground, and there are plenty of signals that public/political impatience with the pace of the transition to a low-GHG economy is growing. We would point to recent decisions by regulators in US states such as California and Indiana, where utilities were urged to reconsider plans to invest in gas-fired generation capacity, and instead to invest in renewable energy with electricity storage. We would also point to calls at the city level to ban natural gas altogether, as happened in Berkeley, CA, or Amsterdam, the Netherlands. In the UK, in mid-2019, the Conservative government at the time decreed that new houses built after 2025 would not be connected to the gas network. Companies may also not be fully convinced of the long-term merits of natural gas. The CEO of Maersk, the world's largest shipping company, publicly challenged his staff to make the company's fleet carbon neutral by 2050. Implicitly, he seemed to suggest that existing technologies, including LNG, are not believed to be up to that task.

In emerging economies, where the overwhelming majority of natural gas demand growth is forecast to take place, the key driver is often not price, or climate change policy, but cleaner air. This narrative is mostly lost in commentaries in OECD countries. Contrary to its possible role in transition pathways towards low-GHG economies, the benefits of natural gas in terms of local air pollutants are undisputed. We discussed the case of China, where in recent years a top-down effort to curtail the use of coal for industrial and district heating in favour of natural gas, plus bans on diesel-fuelled trucks entering major urban areas in the east of the country, incentivized double-digit growth figures for a number of years, and potentially robust growth in the years ahead. The key uncertainty here is whether or not domestic gas production in China can grow to constrain growing import dependence. If not, then geopolitical concerns may restrict the pace of the dash for gas. We speculated that, with the rise of middle classes in many countries in South and Southeast Asia, we may see more governments wanting to address local air pollution, with opportunities for renewable energy and also natural gas. The IEA anticipates that natural gas demand in industry will be the key driver of demand growth in Southeast Asia, trumping electricity generation. The metrics that determine the success, or failure, of natural gas in emerging economies are different, we observed. Here the fuel will have to compete with alternatives on price as much as possible, which is not easy given the widespread availability of more affordable competitors, such as coal in electricity generation, and the fact that a substantial share of natural gas has to be imported as LNG by ship. Market reform to facilitate competition will be an important feature that can help create an environment in which natural gas may succeed. Technology to help distribute natural gas will be critically important, such as the rise of FSRUs, and initial usage of even smaller-scale mini FSRUs, and ISO containers. New business models, like the LNG-to-power project of Total and Siemens in Myanmar, can help unlock new demand. Policy designed to improve local air quality, and achieve health benefits, will continue to be important as well. Modest projects with lower capital costs and short-term pay-off also reduce the danger of lock-in and stranded assets, but lots of small projects can add up to significant demand for producers.

All these trends blur our understanding of natural gas demand, as is reflected in the significant differences in demand forecasts and scenarios, of which we have provided an overview. There is little doubt that natural gas will be an important fuel in the short and medium terms – yet, the further we look ahead, the larger the uncertainties grow. In mature markets, the uncertainties mostly revolve around the question of how substantial the decline in demand in the long term (2040 and beyond) will be. These uncertainties about natural gas demand also explain why investors on the supply side have focused on developing lower-cost resources. Regardless of which scenario turns out to be right, we believe that a third revolution, possibly along the lines we described, will be necessary. Business as usual is not an option for the natural gas industry – or for any industry for that matter, in the context of an accelerating climate crisis. Some producers are taking initial steps to help spur a third revolution, such as voluntary initiatives to improve measurement and reduction of fugitive methane emissions, reduce flaring, and invest in the commercialization of

CCUS technologies. However, much room for improvement remains, and we suggest that the window for action may be limited in duration.

Is this all, then, a matter of market functioning, pricing and technological development? Quite the contrary, we have observed, as we seem to have entered an era of geopolitical rivalry. The impact on natural gas markets may be profound. It is evident that the trade war between China and the United States has consequences: for US LNG project developers that need long-term offtake agreements to secure their loans, for example, not being able to sign contracts in China is bad news. There may be some opportunity in an increasingly commodified LNG market where intermediaries sell on the LNG and its country of origin is no longer apparent. Trade disputes between the US and other countries resulted in steel tariffs, which can drive up the costs of LNG projects in general by as much as 10 per cent. In parts of the EU, the liberal market paradigm is called into question, as the prominence of Russia's Gazprom persists despite the deterioration of EU–Russia relations, and a political desire for more diverse gas supplies. Mounting environmental concerns in parts of the United States and the EU – albeit linked to seismicity, fugitive methane, or more general impatience with the pace (or lack thereof) of GHG emissions reductions – have tarnished the public image of natural gas as a fuel for the future, a trend that may well gain more traction in the years ahead.

Currently, we do not know whether this Golden Age, if that is what it is, will last. We have no doubt that, globally, natural gas will continue to grow in the near and medium term (the 2020s and 2030s), and will have measurable impacts, in terms of improved local air quality and also carbon emissions reductions as a consequence of fuel switching in the power sector. However, if the global community is to make good on its promise to stem the worst impacts of climate change, a third revolution is required for natural gas to play a role in scenarios of deep decarbonization, in 2035 and beyond. Time will tell whether we will witness another revolution, or *hubris* will win out, and the Golden Age proves to be short-lived.

Postscript

As our manuscript goes to press in late March of 2020, the world is grappling with two major events that may end up reshaping global economies, including natural gas markets, for many years to come. First, in December 2019 in Hubei province in China, the COVID-19 virus – or 'Corona virus' in the popular lexicon – started to spread, reaching literally every corner of the world in the timespan of merely three months. Second, on 6 March 2020, negotiations between a group of major oil producers referred to as 'OPEC+' collapsed, unleashing an oil price war. Consequently, just as travel restrictions and lockdowns to combat the spread of the virus dramatically reduced demand for energy, so key oil producers vowed to flood an already oversupplied market with more supplies as they argued over production quota. At the time of writing, both situations are very fluid, and it would be foolish to make bald predictions about the consequences of this unprecedented convergence of crises, for either the short or the longer term. However, we are grateful for the opportunity offered by Polity Press to share some thoughts on what may be ahead for natural gas markets coming out of this pandemic and oil price war.

It is important to note that, even before the situation in China became a global pandemic, as we have noted, natural gas prices were depressed as supply had outpaced demand. However, the extent to which countries around the world are successful in containing the COVID-19 virus will greatly determine demand for energy, including natural gas. Optimists will point to countries such as China and the Republic of Korea, where dramatic restrictions on the movement of people brought the respective economies of these countries close to a standstill with significant downward impact on energy demand, to start rebounding slowly a couple of months later as the number of new confirmed cases declined, and restrictions were slowly loosened. As we write, it is too early for celebration, though, as many questions remain about the longevity of this observed rebound, the extent to which the COVID-19 virus may start to spread again now that public life is picking up, and, specifically in the case of China, how reliable the published data are to begin with. In today's interconnected world, we also cannot help but think that, even if the recovery in East Asia continues, it will undoubtedly be constrained by the dramatic slowdown in economic activity as observed in virtually all other parts of the world where the virus is now spreading. At the time of writing, the World Health Organization has declared Europe as the new epicentre of the pandemic, and the limited data we have tell us that the virus is also spreading rapidly throughout the Americas. As travel is severely restricted, and most factories and stores are forced to close, demand for all fossil fuels, including natural gas, has plummeted. The International Energy Agency has noted that, with large swaths of people working from home, these dire times provide us with a gentle reminder how important stable electricity supply, including generation by burning natural gas, is.[1] At the same time, it is increasingly evident that the COVID-19 virus is forcing various countries around the world into deep economic recession, which will restrain energy demand beyond the short term as well, and negatively impact investment plans to develop new reserves.[2]

It is too early to say whether the current oil price war will be long-lived, or not. Anecdotal evidence suggests that Russian decision-makers had concluded that their three-year experiment in cooperating with OPEC had not been to their benefit, and consequently had no interest in further production cuts. In addition, some influential Russian oil producers figured that they could well benefit from the initial Chinese demand recovery as the country slowly lifted its travel restrictions. Similarly, anecdotal evidence suggests that the rulers of the Kingdom of Saudi Arabia were so frustrated by that message that, instead of maintaining the status quo, they decided that they would rather produce oil at maximum levels in the coming months to fight for market share. The announcement of a price war starting in April has sent Brent oil prices tumbling, from close to $65 in January of this year to below $25 in the third week of March. The consequences will likely be profound, in particular for relatively high-cost producers around the world, including some of the producers of US tight oil, Canadian tar sands, and Venezuelan crude. The consequences will be equally grave for producers with high fiscal break-even prices, such as Iran, Algeria, Nigeria, Iraq and Angola. We are already witnessing large-scale layoffs in the producing and servicing sectors, and will consequently see oil production fall, as soon as storage options have been utilized and oil has nowhere else to go, until meaningful demand recovery that might incentivize drilling to pick up again. For some of the shale gas producers, who have been struggling in a market that has been oversupplied for some time, a drop in tight oil and associated gas production may in fact bring some relief, but it is too early to tell for sure. Making new investment decisions in LNG liquefaction capacity is increasingly complicated, as shown by the rumoured delay of the final investment decision of Mozambique LNG, a project led by Exxon Mobil.[3] Previous price collapses have taught us that US shale will not disappear, but that the industry may well look markedly different from how it does today – quite likely more consolidated – and that the financial community may have a different appetite for this industry, and different metrics on how to assess the value of these resources going forward.

Significant and highly impactful short-term consequences notwithstanding, this leads us to the most pressing longer-term questions that are worth contemplating. The collapse of economic activity, first in China and later in Europe, has unsurprisingly had a major downward effect on greenhouse gas emissions and local air pollution, as factories closed, refineries, steel plants and power plants were curtailed, and travel suspended.[4] One might be tempted, then, to suggest that at least the environment is a net beneficiary of COVID-19, but that would truly be short-sighted; rather, previous economic downturns have shown that environmental and climate policy are at high risk of becoming second- or third-tier policy concerns as governments attempt to restart their economies, rather than the first-tier concern that it has to be given the imperative for significant action to mitigate climate change. An early analysis of China's policy priorities suggests that this time may be no different.[5] A Polish deputy minister opportunistically floated the idea of scrapping the EU emissions trading scheme, or exempting Poland, to combat the virus.[6] Should we then look at COVID-19 responses, as varied as they have been, to identify possible strategies to address climate change? With hundreds of millions of people confined to their homes, and dozens of millions of people about to lose their job due to the economic collapse, further amplifying their uncertainty, it seems highly questionable to make that argument.[7] Time and civil obedience seem required for political leaders and the health industry to contain effectively and, in due course, bring to an end the global health emergency presented by the COVID-19 outbreak. We can also hope that the attention of politicians and industry leaders soon thereafter returns to the urgent concerns raised by climate change.

Notes

1 Natural Gas Fundamentals

1 Oil Change International (2019), *Burning the Gas 'Bridge Fuel Myth': Why Gas Is Not Clean, Cheap or Necessary*. Washington, DC: Oil Change International: http://priceofoil.org/2019/05/30/gas-is-not-a-bridge-fuel.

2 For a comprehensive analysis of these issues, see IEA (2019), *The Role of Gas in Today's Energy Transitions*. Paris: IEA: https://web store.iea.org/the-role-of-gas-in-todays-energy-transitions.

3 For a discussion of the need for a clearer understanding of the political dimensions of natural gas, see B. Shaffer (2013), 'Natural gas supply stability and foreign policy', *Energy Policy*, 56: 114–25.

4 Peter C. Evans and Michael. F. Farina (2013), *The Age of Gas & The Power Networks*: www.ge.com/sites/default/files/GE_Age_of_Gas_Whitepaper_20131014v2.pdf.

5 A. J. Kidnay, W. R. Parrish and D. G. McCartney (2011), *Fundamentals of Natural Gas Processing*, 2nd edn. Florida: CRC Press, p. 135.

6 For an elaborate overview of the basics of natural gas, see V. Chandra (2017), *Fundamentals of Natural Gas – An International Perspective*. Tulsa, OK: PennWell Publishers.

7 E. R. Braziel (2016), *The Domino Effect: How the Shale Revolution Is Transforming Energy Markets, Industries, and Economies*. Houston: NTA Press.

8 The various reports from the IPCC can be found on their website at: www.ipcc.ch.

9 Details of the GGFR Partnership and various reports and country case studies can be found at: www.worldbank.org/en/programs/gas flaringreduction#7.

10 See M. D. Bazilian and J. W. Busby (2019), 'The United States' Gas

flare-up', *Foreign Policy*, 28 July: https://foreignpolicy.com/2019/07/28/the-united-states-gas-flare-up.

11 For an analysis of this issue, see C. A. S. Hall, J. G. Lambert and S. B. Balogh (2014), 'EROI of different fuels and the implications for society', *Energy Policy*, 64: 141–52.

12 Ajay Makan and Ed Crooks (2013), 'Shale Gas Boom now visible from space', *Financial Times*, 27 January: www.ft.com/content/d2d2e83c-6721-11e2-a805-00144feab49a.

13 Bloomberg New Energy Finance (2018), 'Permian Basin flaring: the view from above', 30 October.

14 Kevin Crowley and Ryan Collins (2019), 'Oil producers are burning enough "waste" gas to power every home in Texas', Bloomberg, 10 April: www.bloomberg.com/news/articles/2019-04-10/permian-basin-is-flaring-more-gas-than-texas-residents-use-daily.

15 US Geological Survey (2001), *Natural Gas Hydrates: Vast Resource, Uncertain Future*. Fact sheet FS-021-01: https://pubs.usgs.gov/fs/fs021-01/fs021-01.pdf.

16 For a more detailed analysis of tight oil development economics, see R. L. Kleinberg, S. Paltsev, C. K. Ebinger, D. A. Hobbs and T. Boersma (2018), 'Tight oil market dynamics: benchmarks, breakeven points, and inelasticities', *Energy Economics*, 70: 70–83.

17 For a discussion of global production cost curves, see R. F. Aguilera (2014), 'Production costs of global conventional and unconventional petroleum', *Energy Policy*, 64: 134–40.

18 T. Mitrova and T. Boersma (2018), *The Impact of US LNG on Russian Natural Gas Export Policy*. New York: Center on Global Energy Policy, SIPA, Columbia University: https://energypolicy.columbia.edu/sites/default/files/pictures/Gazprom%20vs%20US%20LNG_CGEP_Report_121418_2.pdf.

19 *Financial Times* (2019), 'Nord Stream 2: stream of cost-consciousness', 13 February: www.ft.com/content/be550730-2f86-11e9-8744-e7016697f225.

20 The International Gas Union has published a *Natural Gas Conversion Guide* (2012) that is available at http://members.igu.org/old/gas-knowhow/Natural%20Gas%20Conversion%20Guide.pdf/view, and the IEA has created an online converter that is available at www.iea.org/statistics/resources/unitconverter.

21 The various publications of the IEA can be purchased from their website, www.iea.org, while the publications of BP are freely available

at www.bp.com/en/global/corporate/about-bp/energy-economics/sta tistical-review-of-world-energy.html, and publications of the EIA are available at www.eia.gov.

22 The main report – *Bowland Shale Gas Study* (2013) – and various appendices can be found at www.gov.uk/government/publications/ bowland-shale-gas-study.

23 Details of the EIA's surveys can be found at www.eia.gov/analysis/ studies/worldshalegas.

24 The cost estimates used here and below for LNG are Evans and Farina, *The Age of Gas,* 30-13. Available at www.ge.com/sites/ default/files/GE_Age_of_Gas_Whitepaper_20131014v2.pdf.

25 The carbon intensity of different sources of natural gas was examined in the IEA's *World Energy Outlook 2017*, which is available at www.iea.org/weo2017.

26 The difference between LNG and CNG is that LNG is cooled and stored under low pressure and is economical to transport across long distances, while CNG is not cooled, is stored under high pressure and is not economical to transport over long distances.

27 N. Tsafos (2018), *Is Gas Global Yet?* Center for Strategic and International Studies commentary. Available at www.csis.org/analy sis/gas-global-yet.

28 The data for 2018 and 2008 were obtained from various issues of the *BP Statistical Review of World Energy*, London: BP.

29 For a nuanced discussion about the analogy between oil price discovery and the market for LNG, see Fulwood, *Asian LNG Trading Hub: Myth or Reality?* New York: Center on Global Energy Policy, Columbia University. Available at https://energypolicy.columbia.edu/ sites/default/files/pictures/Asian%20LNG%20Trading%20Hubs_ CGEP_Report_050318.pdf.

30 For background analysis, see J. P. Stern (ed.) (2012), *The Pricing of Internationally Traded Gas*. Oxford: OIES / Oxford University Press.

31 See Fulwood, *Asian LNG Trading Hub*.

32 N. Tsafos (2019), *Gas Line, Q2 2019*. Washington, DC: Center for Strategic and International Studies commentary. Available at www. csis.org/analysis/gas-line-q2-2019.

33 Mitrova and Boersma, *The Impact of US LNG on Russian Natural Gas Export Policy*.

34 There are now a large number of handbooks on various aspects of energy security, and the majority tend to assume that oil and gas

security are similar – very few have chapters dedicated to natural gas. Boersma gives an overview of energy security literature as it relates to natural gas in the EU, and observes an overwhelming focus in the literature on market functioning and conceptual contributions, see: T. Boersma (2015), *Energy Security and Natural Gas Markets in Europe: Lessons from the EU and the United States*, Series in Energy Policy. London: Routledge.

35 T. Mitrova, T. Boersma and A. Galkina (2016), 'Some future scenarios of Russian gas in Europe', *Energy Strategy Reviews*, 11–12: 19–28.

36 A. Goldthau and T. Boersma (2014), 'The 2014 Ukraine–Russia crisis: implications for energy markets and scholarship', *Energy Research & Social Science*, 3: 13–15.

37 A. Halff, L. Younes and T. Boersma (2019), 'The likely implications of the new IMO standards on the shipping industry', *Energy Policy*, 126: 277–86.

38 J. Bordoff and N. Kaufman (2018), 'A federal US carbon tax: major design decisions and implications', *Joule*, 2 (12): 2487–91.

39 J. Stern (2017), *Challenges to the Future of Gas: Unburnable or Unaffordable?* NG 125. Oxford: OIES. Available at www.oxford energy.org/wpcms/wp-content/uploads/2017/12/Challenges-to-the-Future-of-Gas-unburnable-or-unaffordable-NG-125.pdf.

40 US Department of Energy (2019), *Department of Energy Authorizes Additional LNG Exports from Freeport LNG*, 29 May. Available at www.energy.gov/articles/department-energy-authorizes-additional-lng-exports-freeport-lng.

2 Pipeline Geopolitics

1 This refers to reports by BP – see, for example, www.naturalgasi ntel.com/articles/101394-lng-to-dominate-global-natgas-trade-by-2035-says-bp.

2 Various analyses use different definitions of 'Europe'. The IEA considers Turkey to be part of the European gas market in most of its reports, even though the Turkish market has limited physical connection, and modest overall resemblance, to the EU market model.

3 All of these statistics are drawn from the 2009 and 2019 editions of the *BP Statistical Review of World Energy*, London: BP.

4 Richard E. Ericson (2009), 'Eurasian natural gas pipelines: the political economy of network interdependence', *Eurasian Geography and Economics*, 50 (1): 28–57.

5 D. Yergin (2006). 'Ensuring energy security', *Foreign Affairs*, March/April: www.foreignaffairs.com/articles/2006-03-01/ensuring-energy-security.

6 For relevant news coverage, see www.bbc.com/news/av/world-europe-44793764/trump-germany-is-totally-controlled-by-russia.

7 For a historical account of the origins of natural gas trade between the Soviet Union and various European countries, see P. Högselius (2012), *Red Gas: Russia and the Origins of European Energy Dependence*. Basingstoke: Palgrave MacMillan, pp. 45–63.

8 J. Elkind and T. Boersma (2018), *Talking Past Each Other: Transatlantic Perspectives on European Gas Security*. New York: Center on Global Energy, Policy Commentary, Columbia University. Available at https://energypolicy.columbia.edu/sites/default/files/pictures/TalkingPastEach%20Other_CGEP_FINAL.pdf.

9 See, for example, European Commission communication on a European Energy Security Strategy, SWD(2014) 330 final:3. Available at www.eesc.europa.eu/resources/docs/european-energy-security-strategy.pdf.

10 For an early exploration of this divergence between the EU and Russia, see S. Boussena and C. Locatelli (2013), 'Energy institutional and organizational changes in EU and Russia: revisiting gas relations', *Energy Policy*, 55: 180–9.

11 For the binding obligation, see https://europa.eu/rapid/press-release_IP-18-3921_en.htm.

12 S. Pirani, J. Stern and K. Yafimava (2009), *The Russo-Ukrainian Gas Dispute of January 2009: A Comprehensive Assessment*, NG 27. Oxford: OIES and Oxford University Press. Available at www.oxfordenergy.org/publications/the-russo-ukrainian-gas-dispute-of-january-2009-a-comprehensive-assessment/?v=3e8d115eb4b3.

13 Minister Sikorski made this analogy speaking at a transatlantic security conference on 30 April 2006. For press coverage of the event, see https://euobserver.com/foreign/21486.

14 See, for example, press coverage from the Nord Stream company: www.nord-stream.com/press-info/press-releases/nord-stream-reaches-average-utilisation-of-93-in-2017-51-bcm-delivered-to-the-european-union-500.

15 To illustrate this, see European Commission Vice President Sefcovic's speech from 2015: https://europa.eu/rapid/press-release_SPEECH-15-4918_en.htm.

16 Mark Smedley (2017), 'PGNIG doubles Qatargas contract', *Natural Gas World*, 14 March, www.naturalgasworld.com/pgnig-doubles-qatargas-contract-36411?#signin.

17 In this interview with the Aspen Review, Czech Ambassador Bartuska states that he does not believe in small regional groupings, but rather in European cooperation in its entirety: www.aspenreview.com/article/2018/vaclav-bartuska-perspective-world-determined-source-raw-materials-energy-supplies.

18 For a comprehensive overview of natural gas infrastructure investments, see K. Yafimava (2018), *Building New Gas Transportation Infrastructure in the EU – What Are the Rules of the Road?* NG 134. Oxford: OIES and Oxford University Press. Available at www.oxfordenergy.org/wpcms/wp-content/uploads/2018/07/Building-New-Gas-Transportation-Infrastructure-in-the-EU-what-are-the-rules-of-the-game.pdf?v=7516fd43adaa.

19 T. Boersma, T. Mitrova, G. Greving and A. Galkina (2014), *Business as Usual: European Gas Market Functioning in Times of Turmoil and Increasing Import Dependence*, Policy Brief 14-05. Washington, DC: Brookings Institution. Available at www.brookings.edu/wp-content/uploads/2016/07/business_as_usual_final_3.pdf.

20 The press release for this historical decision can be found at https://acer.europa.eu/Media/News/Pages/ACER-adopts-a-decision-on-the-allocation-of-costs-for-the-Gas-Interconnection-project-between-Poland-and-Lithuania.aspx.

21 See, for example, www.euractiv.com/section/energy/news/energy-union-completed-as-commissions-final-stocktake-debuts.

22 The communication on these stress tests, plus supporting documents and analyses, can be found at https://ec.europa.eu/energy/en/news/stress-tests-cooperation-key-coping-potential-gas-disruption.

23 The press release on the stress tests confirms, e.g., that the EU energy system is reasonably resilient to supply shocks and that market mechanisms tend to function, and in dire times can be supported by non-market mechanisms. Nowhere in this document is the conclusion that volumes of Russian gas imported have to be reduced; however, this did become a cornerstone of policy initiatives in the years after 2014: https://europa.eu/rapid/press-release_MEMO-14-593_en.htm.

24 For more details, see S. Pirani (2018), *Russian Gas Transit through Ukraine after 2019: The Options*, Oxford Energy Insight 41. Oxford: OIES and Oxford University Press. Available at www.oxfordener gy.org/wpcms/wp-content/uploads/2018/11/Russian-gas-transit-th rough-Ukraine-after-2019-Insight-41.pdf.

25 T. Mitrova, T. Boersma and A. Galkina (2016), 'Some future sce narios for Russian natural gas in Europe', *Energy Strategy Reviews*, 11–12: 19–28.

26 See, for instance, www.reuters.com/article/us-eu-gazprom-us-idUSKCN0XX1YG.

27 Both Energy Secretary Perry and President Trump repeatedly referenced US LNG while discussing the risks of constructing Nord Stream 2; see, for instance, www.dw.com/en/us-to-eu-our-liquefied-natural-gas-is-more-reliable-than-russias/a-48576208.

28 T. Boersma and C. M. Johnson (2018), *US Energy Diplomacy*. New York: Center on Global Energy Policy, Columbia University. Available at https://energypolicy.columbia.edu/sites/default/files/pict ures/CGEPUSEnergyDiplomacy218.pdf.

29 In 2018, Gazprom imports into the EU reached their highest level on record, see www.reuters.com/article/us-gazprom-results/record-russ ian-gas-sales-to-europe-help-gazprom-profits-double-idUSKCN1S 51DU.

30 For an extensive analysis of these issues, see Andreas Goldthau and Nick Sitter (2015), *A Liberal Actor in a Realist World: The European Union Regulatory State and the Global Political Economy of Energy*. Oxford University Press.

31 For a reassessment of Central Asia's relative location, see Kent E. Calder (2012), *The New Continentalism: Energy and Twenty-First-Century Eurasian Geopolitics*. New Haven: Yale University Press, pp. 37–8.

32 For a detailed examination of the evolution of the Eurasian pipeline system, see Andreas Heinrich (2014), 'Introduction: export pipelines in Eurasia'. In Andreas Heinrich and Heiko Pleines (eds.), *Export Pipelines from the CIS Region: Geopolitics, Securitization and Political Decision-Making*. Stuttgart: Ibidem-Verlag, pp. 13–73.

33 Boris Barkanov (2014), 'The geo-economics of Eurasian gas: the evolution of Russia–Turkmen relations in natural gas (1992–2010)', in Heinrich and Pleines (eds.), *Export Pipelines from the CIS Region*, 149–74.

34 Farkhod Aminjonov (2018), 'Central Asian gas export dependency: swapping Russian patronage for Chinese', *RUSI Journal*, April/May, 163 (2): 66–77.

35 Slavomir Horak (2012), 'Turkmenistan's shifting energy geopolitics in 2009–2011: European perspectives', *Problems of Post-Communism*, 59 (2): 18–30.

36 A. Grigas (2017), *The New Geopolitics of Natural Gas*. Cambridge, MA: Harvard University Press.

37 For a detailed analysis of Western involvement and the deals with China, see Annette Bohr (2016), *Turkmenistan: Power, Politics and Petro-Authoritarianism*. London: Chatham House. Available at www.chathamhouse.org/sites/default/files/publications/research/20 16-03-08-turkmenistan-bohr.pdf.

38 See https://uk.reuters.com/article/gas-turkmenistan-china/turkmen-gas-exports-to-china-to-hit-65-bcm-year-by-2020-idUKL6N0GZ0 OT20130903.

39 Alex Forbes (2019), 'Turkmenistan sees light at the end of the tunnel', *Petroleum Economist*, 10 June: www.petroleum-economist. com/articles/upstream/exploration-production/2019/turkmenistan-sees-light-at-the-end-of-the-tunnel.

40 See Bohr, *Turkmenistan*, 78.

41 *Bloomberg News* (2018), 'China's traditional gas allies fail to meet demand boom in Winter', 25 February: www.bloomberg.com/news/articles/2018-02-25/china-s-traditional-gas-allies-fail-to-meet-winter-demand-boom.

42 Yusin Lee (2017), 'Turkmenistan's East–West Gas Pipeline: will it save the country from economic decline?' *Problems of Post-Communism*: https://doi.org/10.1080/10758216.2017.1366273.

43 Olga Hryniuk (2018), 'The escalation of Iran–Turkmenistan gas dispute: will the battle begin?' *CIS Arbitration Forum*, 6 March. Available at www.cisarbitration.com/2018/03/06/the-escalation-of-iran-turkmenistan-gas-dispute-will-the-battle-begin.

44 Interfax Global Energy (2019), 'Azerbaijan stops gas imports with none planned for 2019', 19 November: http://interfaxenergy.com/article/33349/azerbaijan-stops-gas-imports-with-none-planned-for-2019.

45 For a discussion of TAPI, see Luca Anceschi (2017), 'Turkmenistan and the virtual politics of Eurasian energy: the case of the TAPI pipeline project', *Central Asian Survey*, 36 (4): 409–29.

46 R. E. Ericson (2013), 'Eurasian natural gas: significance and recent developments', *Eurasian Geography and Economics*, 53 (5): 615–48.

47 For a comprehensive analysis, see Simon Pirani (2018), *Let's Not Exaggerate: Southern Gas Corridor Prospects to 2030*, NG 135. Oxford: OIES and Oxford University Press. Available at www.oxfordenergy.org/publications/lets-not-exaggerate-southern-gas-cor ridor-prospects-2030/?v=79cba1185463.

48 Carla P. Freeman (2018), 'New strategies for an old rivalry? China–Russia relations in Central Asia after the energy boom', *The Pacific Review*, 31 (5): 635–54.

49 *Oxford Analytica Daily Brief* (2019), 'Turkmenistan short of gas field and export funding', 4 September.

50 Matthew Sagers (1995), 'Prospects for oil and gas developments in Sakhalin Oblast', *Post-Soviet Geography*, 47 (5): 505–45.

51 Michael Bradshaw (2010), 'A new energy age in Pacific Russia: lessons from the Sakhalin oil and gas projects', *Eurasian Geography and Economics*, 51 (3): 330–59.

52 Bureau of Industry and Security (2015), *Russian Sanctions: Addition to the Entity List to Prevent Violations of Russian Industry Sector Sanctions*, 7 August. Available at www.federalregister.gov/docum ents/2015/08/07/2015-19274/russian-sanctions-addition-to-the-enti ty-list-to-prevent-violations-of-russian-industry-sector.

53 For an encyclopaedic analysis of Sino-Russian energy relations, see Keun-Wook Paik (2012), *Sino-Russian Oil and Gas Cooperation: The Reality and Implications*. Oxford: OIES and Oxford University Press.

54 Jonathan Stern and Michael Bradshaw (2008), 'Russian and Central Asian gas supply for Asia'. In Jonathan Stern (ed.), *Natural Gas in Asia: The Challenges of Growth in China, India, Japan and Korea*, 2nd edn. Oxford: OIES and Oxford University Press, pp. 220–78.

55 Sarah Lain (2015), 'The Bear and the Dragon', *RUSI Journal*, 160 (1): 68–67.

56 Keun-Wook Paik (2015), *Sino-Russian Gas and Oil Cooperation: Entering a New Era of Strategic Partnership?* WPM 59. Oxford: OIES and Oxford University Press. Available at www.oxfordenergy.org/wpcms/wp-content/uploads/2015/04/WPM-59.pdf.

57 For an analysis of the changing bargaining position and interests between Russia and China, see Tom Roseth (2017), 'Russia's energy

relations with China: crossing the strategic threshold', *Eurasian Geography and Economics,* 58 (1): 23–55.

58 For a discussion of the Altai Pipeline, see James Henderson (2014), *The Commercial and Political Logic for the Altai Pipeline,* Oxford Energy Comment. Oxford: OIES and Oxford University Press. Available at www.oxfordenergy.org/wpcms/wp-content/uploads/20 14/12/The-Commercial-and-Political-Logic-for-the-Altai-Pipeline-GPC-4.pdf.

59 For an assessment of Russia's Asia Gas Pivot, see James Henderson (2018), *Russia's Gas Pivot to Asia: Another False Dawn or Ready to Lift Off?* Oxford Energy Insight 40. Oxford: OEIS.

60 Y. Lee (2019), 'Political viability of the Russia – North Korea – South Korea Gas Pipeline project: an analysis of the role of the U.S.', *Energies,* 12: https://doi.org/10.3390/en12101895.

61 IEA (2018), *World Energy Outlook 2018.* Paris: IEA, p. 171.

62 D. Sandalow, A. Losz and S. Yan (2018), *A Natural Gas Giant Awakens: China's Quest for Blue Skies Shapes Global Markets.* Columbia University: Center on Global Energy Policy. Available at https://energypolicy.columbia.edu/sites/default/files/file-uploads/ China%20Nat%20Gas%20Commentary_CGEP_June%202018_ FINAL.pdf.

3 THE SHALE GAS REVOLUTION

1 EIA (2019), 'United States has been a net exporter of natural gas for more than 12 consecutive months', *Today in Energy,* 2 May: www. eia.gov/todayinenergy/detail.php?id=39312.

2 BP (2019), *The US Energy Market in 2018.* London: BP. Available at www.bp.com/en/global/corporate/energy-economics/statistical-review-of-world-energy/country-and-regional-insights/united-states. html.

3 T. Boersma and C. Johnson (2013), *Shale Gas in Europe: A Multidisciplinary Analysis with a Focus on European Specificities,* European Energy Studies 5. Deventer, The Netherlands: Claeys & Casteels Publishers.

4 For a more detailed overview of this terminology, see T. Boersma and P. Andrews-Speed (2018). In R. Bleischwitz (ed.), *Routledge Handbook of the Resource Nexus.* London: Routledge.

5 www.energy.gov/sites/prod/files/2015/05/f22/QER%20Analysis%
20-%20Opportunities%20for%20Efficiency%20Improvements%
20in%20the%20U.S.%20Natural%20Gas%20Transmission%20
Storage%20and%20Distribution%20System.pdf.

6 We base this number on Federal Energy Regulatory Commission
(FERC) data of approved major pipeline projects, and assume that,
once approved, these projects are completed. We have not included
the ten projects that were approved just prior to the summer of 2019.
See www.ferc.gov/industries/gas/indus-act/pipelines/approved-proje
cts.asp.

7 K. B. I. Medlock (2012), 'Modeling the implications of expanded US
shale gas production', *Energy Strategy Reviews*, 1: 33–41.

8 The EIA provides detailed information on natural gas at the state
level. Available at www.eia.gov/dnav/ng/ng_prod_sum_a_EPG0_
FPD_mmcf_a.htm.

9 There is a substantial, and growing, body of social science research
on public attitudes to shale gas in the US and Canada – see, for example, M. Thomas et al. (2017), 'Public perceptions of hydraulic fracturing for shale gas and oil in the United States and Canada', *Wiley
Interdisciplinary Reviews: Climate Change*, 8 (3): 345e (https://doi.
org/10.1002/wcc.450).

10 For media reports, see www.bostonglobe.com/business/2018/01/29/
tanker-unloads-lng-everett-terminal-that-contains-russian-gas/rewj
1wKjajaKtLp79irzTI/story.html.

11 www.nytimes.com/2014/12/18/nyregion/cuomo-to-ban-fracking-in-
new-york-state-citing-health-risks.html.

12 For the 2019 root cause analysis of the Aliso Canyon leak, see ftp://
ftp.cpuc.ca.gov/News_and_Outreach/SS-25%20RCA%20Final%20
Report%20May%2016,%202019.pdf.

13 California state regulators have urged utilities to reconsider planned
investments in gas-fired generation capacity in favour of renewables plus storage technologies, see www.greentechmedia.com/
articles/read/natural-gas-under-threat-california-pge-gas-plants-
energy-storage#gs.wyu4qa.

14 The Baker Hughes Rig Count website gives a detailed accounting of
the number of rigs in operation on a weekly basis in Canada, the US
and internationally: https://bakerhughesrigcount.gcs-web.com/rig-
count-overview.

15 For a thorough analysis of the role of the US government in the

development of hydraulic fracturing in shale, see A. Trembath, J. Jenkins, T. Nordhaus and M. Shellenberger (2012), *Where the Shale Gas Revolution Came From*. Oakland, CA: Breakthrough Institute. Available at www.ourenergypolicy.org/wp-content/uploads/2012/05/Where_the_Shale_Gas_Revolution_Came_From.pdf.

16 D. Raimi (2018), *The Fracking Debate – The Risks, Benefits, and Uncertainties of the Shale Revolution*. New York: Columbia University Press.

17 Emily Pickrell (2014), 'Economist: shale fever soon will decline', *Fuel Fix*, 25 February: https://fuelfix.com/blog/2014/02/25/economist-shale-fever-soon-will-decline.

18 Paul Stevens has written a number of influential reports on shale gas for Chatham House – see Stevens (2010), *'The Shale Gas Revolution': Hype and Reality*. London: Chatham House. Available at www.chathamhouse.org/sites/default/files/field/field_document/r_0910stevens.pdf.

19 Trevor Houser and Shashank Mohan (2014), *Fueling Up – The Economic Implications of America's Oil and Gas Boom*. Washington, DC: Peterson Institute for International Economics.

20 For some background on falling employment numbers in North Dakota in 2015, see www.huduser.gov/portal/pdredge/pdr-edge-spot light-article-032116.html.

21 The origins of this terminology are not entirely clear. J. Stern (2018), 'LNG Market Outlook Europe', LNG Producer–Consumer Conference, Nagoya, 22 October, refers to the EU as a market of last resort for LNG that cannot find a home elsewhere: www.lng-conference.org/presen/pdf/Session3_Stern_LNG.pdf.

22 EIA (2018), 'EIA forecasts natural gas to remain primary energy source for electricity generation', *Today in Energy*, 22 January: www.eia.gov/todayinenergy/detail.php?id=34612.

23 EIA (2019), *Annual Energy Outlook 2019 with Projections to 2050*: www.eia.gov/outlooks/aeo/pdf/aeo2019.pdf.

24 L. Michael Buchsbaum (2019), 'In 2019, expect even less coal in the US', *Energy Transition*, 25 January: https://energytransition.org/2019/01/in-2019-expect-even-less-coal-in-the-us.

25 See www.earth-policy.org/data_highlights/2013/highlights41.

26 EPA (2019), *Inventory of US Greenhouse Gases and Sinks: 1990–2017*: www.epa.gov/sites/production/files/2019-04/documents/us-ghg-inventory-2019-main-text.pdf.

27 R. W. Howarth (2019), 'Ideas and perspectives: is shale gas a major driver of recent increase in global atmospheric methane?' *Biogeosciences*, 16: 3033–46 (https://doi.org/10.5194/bg-16-3033-2019).

28 Clifford Krauss (2019), 'Trump's methane rule rollback divides oil and gas industry', *New York Times*, 29 August: www.nytimes.com/2019/08/29/business/energy-environment/methane-regulation-reaction.html.

29 For an early call for this more holistic approach, see J. R. Pielke (2010), *The Climate Fix – What Scientists and Politicians Won't Tell You About Global Warming*. New York: Basic Books.

30 One widely reported case is in Pavillion, Wyoming, where court cases continue to date. See this Stanford University overview for research findings from 2016: https://news.stanford.edu/2016/03/29/pavillion-fracking-water-032916.

31 For an overview of induced seismicity and links to the oil and gas industry in Oklahoma, see http://theconversation.com/earthquakes-from-the-oil-and-gas-industry-are-plaguing-oklahoma-heres-a-way-to-reduce-them-91044.

32 T. Boersma (2016), *Natural Gas in the United States in 2016: Problem Child and Poster Child*, Policy brief 16-02. Washington, DC: Brookings Institution. Available at www.brookings.edu/research/natural-gas-in-the-united-states-in-2016-problem-child-and-poster-child.

33 See, for example, this 2012 speech, in which President Obama points to increased drilling activity in the United States, while also noting that uncontrolled consumption cannot be a sustainable path forward: https://obamawhitehouse.archives.gov/the-press-office/2012/02/23/remarks-president-energy.

34 The 2019 US Democratic primary debates give a good indication of the anti-fossil fuel sentiment that is growing in the party – see, for instance, www.vox.com/policy-and-politics/2019/7/31/20748784/2020-democratic-debates-climate-change-fossil-fuels.

35 EIA (2019), *Natural Gas Explained: Imports and Exports*: www.eia.gov/energyexplained/index.php?page=natural_gas_imports.

36 Between February 2016 and December 2017, fifty cargos of US LNG were delivered in Mexico, making it the number one destination for US LNG, or 18.8 per cent of total LNG cargos shipped; see www.energy.gov/sites/prod/files/2018/04/f50/LNG%20Annual%20Report%20-%202017.pdf.

37 Cheniere Energy received its licence to export LNG from its Sabine Pass terminal in Louisiana in 2010; for news coverage, see www.wsj. com/articles/SB10001424052748704644404575482290198119482.

38 American Chemistry Council (2019), *U.S. Chemical Investment Linked to Shale Gas: $204 Billion and Counting*, May: www.americ anchemistry.com/Shale_Gas_Fact_Sheet.aspx.

39 For the full text of Secretary Clinton's speech, see https://2009-2017. state.gov/secretary/20092013clinton/rm/2012/10/199330.htm.

40 Meghan L. O'Sullivan (2017), 'US energy diplomacy in an age of energy abundance', *OIES Forum*, 111 (November): 8–1. Available at www.oxfordenergy.org/wpcms/wp-content/uploads/2018/01/OEF-111.pdf. See also Tim Boersma and Corey Johnson (2018), *U.S. Energy Diplomacy*. New York: Columbia / SIPA, Center on Global Energy Policy. Available at https://energypolicy.columbia.edu/sites/ default/files/pictures/CGEPUSEnergyDiplomacy218.pdf.

41 Andreas Goldthau (2018), *The Politics of Shale Gas in Eastern Europe: Energy Security, Contested Technologies and the Social Licence to Frack*. Cambridge University Press, p. 22.

42 EIA (2014), *Technically Recoverable Shale Oil and Shale Gas Resources: An Assessment of 137 Shale Formations in 41 Countries Outside the United States*. Washington, DC: US EIA. Available at www.eia.gov/analysis/studies/worldshalegas/archive/2013/pdf/fullre port_2013.pdf.

43 See https://oilprice.com/Latest-Energy-News/World-News/Argenti na-Exports-First-Ever-Cargo-Of-LNG.html.

44 See chapter 2, 'The policy context'. In Goldthau, *The Politics of Shale Gas in Eastern Europe*, 18-47.

45 IEA (2012), *Golden Rules for a Golden Age of Gas*. Paris: IEA. Available at https://webstore.iea.org/weo-2012-special-report-golden-rules-for-a-golden-age-of-gas. Also European Commission, *COMMISSION RECOMMENDATION of 22 January 2014 on Minimum Principles for the Exploration and Production of Hydrocarbons (such as Shale Gas) Using High-Volume Hydraulic Fracturing*. Brussels: European Commission. Available at https://eur-lex.europa.eu/LexUriServ/LexUriServ.do?uri=OJ:L:2014:039:0072: 0078:EN:PDF.

46 M. H. A. A. Zijp, S. Nelskamp and J. C. Doornenbal (2017), *Resource Estimation of Shale Gas and Shale Oil in Europe*, Report T7b of

the EUOGA (EU Unconventional Oil and Gas Assessment) study commissioned by the European Commission Joint Research Centre. Available at https://ec.europa.eu/jrc/sites/jrcsh/files/t7_resource_estimation_of_shale_gas_and_shale_oil_in_europe.pdf.

47 C. Johnson and T. Boersma (2013), 'Energy (in)security in Poland: the case of shale gas', *Energy Policy*, 53: 389–99 (https://doi.org/10.1016/j.enpol.2012.10.068).

48 This section draws on ibid.; Michael Carnegie Labelle (2017), 'Failure to frack: pitfalls of governance and risk in Polish shale gas', in R. Q. Graffton, G. Cronshaw and M. C. Moor (eds.), *Risk, Rewards and Regulation of Unconventional Gas*. Cambridge University Press, pp. 267–85; and Michael Labelle (2018), 'Poland disappointed expectations: energy security vs. bureaucracy and geology', in Shanti Gamper-Rabindran (ed.), *The Shale Dilemma: A Global Perspective on Fracking and Shale Development*. University of Pittsburgh Press, pp. 178–203.

49 Separate pipeline import data are not available in BP's 2019 statistical review.

50 G. Kemp, C. Johnson and T. Boersma (2012), *The Shale Gas Boom: Why Poland Is Not Ready*. German Marshall Fund of the United States. Available at www.gmfus.org/publications/shale-gas-boom-why-poland-not-ready.

51 Goldthau, *The Politics of Shale Gas in Eastern Europe*.

52 This section draws on Michael Bradshaw (2017), 'Unconventional gas in the United Kingdom'. In Grafton et al. (eds.), *Risk, Rewards and Regulation of Unconventional Gas*, 167–96; and Michael Bradshaw and Catherine Waite (2017), 'Learning from Lancashire: exploring the contours of the shale gas conflict in England', *Global Environmental Change*, 47: 28–36 (https://doi.org/10.1016/j.gloenvcha.2017.08.005).

53 Oil and Gas Authority (2019), *UK Oil and Gas Production and Projections of UK Oil and Gas Production and Expenditure 2018 Report – March 2019*. London: OGA. Available at www.ogauthority.co.uk/media/5382/oga_projections-of-uk-oil-and-gas-production-and-expenditure.pdf.

54 See www.gov.uk/government/news/local-councils-to-receive-millions-in-business-rates-from-shale-gas-developments.

55 For the Ministerial Notice, see www.gov.uk/government/news/written-ministerial-statement-by-edward-davey-exploration-for-shale-gas.

56 Details of the Department of Business, Energy & Industrial Strategy (BEIS) Public Attitudes Tracker can be found at www.gov.uk/gover nment/collections/public-attitudes-tracking-survey.

57 Department of Business, Energy & Industrial Strategy (2019), *UK National Risk Assessment of Security of Gas Supply 2018*. London: BEIS. Available at www.gov.uk/government/publications/uk-natio nal-risk-assessment-on-security-of-gas-supply-2018.

58 For a detailed discussion of China's shale resources, see L. V. Jianzhong and Zhang Huanzhi (2017), 'Unconventional gas in China'. In Grafton et al. (eds.), *Risk, Rewards and Regulation of Unconventional Gas*, 129–41.

59 For early comprehensive assessments of China's shale gas potential, see Fan Gao (2012), *Will There Be a Chinese Shale Gas Revolution in China 2020?* Oxford: OIES and Oxford University Press; and David Sandalow, Jingchao Wu, Qing Yang, Anders Hove and Junda Lin (2014), *Meeting China's Shale Gas Goals*. New York: Center on Global Energy Policy, Columbia/SPA.

60 CNBC (2019), 'China's gas demand growth rate to slow in 2019, government report shows', 2 September: www.cnbc.com/2019/09/02/chinas-gas-demand-growth-rate-to-slow-in-2019-government-re port-shows.html.

61 Lei Tian et al. (2014), 'Stimulation of shale gas development in China: a comparison with the US experience', *Energy Policy*, 75: 109–16 (https://doi.org/10.1016/j.enpol.2014.07.025).

62 J. Chen (2018), 'Shale gas exploration and development progress in China and the way forward', *IOP Conference Series: Earth and Environmental Science*, 113: https://iopscience.iop.org/article/10.1088/1755-1315/113/1/012178/pdf.

63 Wood MacKenzie (2019), *Chinese Shale Gas Production Will Almost Double in Two Years*, 17 April 2018: www.woodmac.com/news/editorial/chinese-shale-gas-production-will-almost-double-in-two-years.

64 Wood Mackenzie (2019), *China's Gas Production to Double to 325 Bcm in 2040*, 28 August 2019: www.woodmac.com/press-releases/chinas-gas-production-to-double-to-325-bcm-in-2040.

65 Alvin Lin (2018), 'Replacing coal with shale gas: could reducing China's regional air pollution lead to more local pollution in rural China?' In Gamper-Rabindran (ed.), *The Shale Dilemma*, 267–304.

66 BP (2019), *The US Energy Market in 2018*. London: BP. Available

at www.bp.com/en/global/corporate/energy-economics/statistical-review-of-world-energy/country-and-regional-insights/united-states. html.

67 For a discussion of the situation in Argentina, see Maria Florencia Saulino (2018), 'Argentina energy extraction and communities: can shale development proceed without causing pollution and conflicts?' In Gamper-Rabindran (ed.), *The Shale Dilemma*, 305–41.

68 See also R. D. Blackwill and J. M. Harris (2016), *War by Other Means – Geoeconomics and Statecraft*. Cambridge, MA: Harvard University Press.

69 IHS Markit has estimated that over 1,250 tcf of natural gas can be produced with Henry Hub prices of 4 $/MMBTU or lower: www. ogj.com/drilling-production/production-operations/unconventional-resources/article/17296611/ihs-markit-us-gas-production-to-rise-60-in-next-20-years.

70 R. L. Kleinberg, S. Paltsev, C. K. Ebinger, D. A. Hobbs and T. Boersma (2018), 'Tight oil market dynamics: benchmarks, breakeven points, and inelasticities', *Energy Economics*, 70: 70–83.

4 The Coming of Age of LNG

1 G. J. Bridge and M. Bradshaw (2017), 'Making a global gas market: territoriality and production networks in Liquefied Natural Gas', *Economic Geography*, 93 (3): 215–40.

2 M. D. Tusiani and G. Shearer (2016), *LNG: Fuel for a Changing World – A Nontechnical Guide*, 2nd edition. Tulsa, OK: PennWell Publishers.

3 A. Aïssaoui (2001), *Algeria: The Political Economy of Oil and Gas*. Oxford: OIES and Oxford University Press.

4 Tusiani and Shearer, *LNG*.

5 BP (2009), *BP Statistical Review of World Energy, June 2009*. London: BP, p. 30.

6 GIIGNL (2019), *The LNG Industry 2019*. Neuilly-Sur-Seine: GIIGNL. Available at https://giignl.org/sites/default/files/PUBLIC_AREA/Publications/giignl_annual_report_2019-compressed.pdf.

7 BP (2019), *BP Statistical Review of World Energy, June 2019*. London: BP, p. 34.

8 Andrew Flower (2011), 'Qatari LNG'. In B. Fattouch and J. Stern

(eds.), *Natural Gas Markets in the Middle East and North Africa*. Oxford: OIES and Oxford University Press, pp. 343–85.

9 For an excellent brief history of Qatar's LNG industry, see K. Hashimoto, J. Elass and S. Eller (2004), *Liquefied Natural Gas from Qatar: The Qatargas Project*. Geopolitics of Gas Study, a joint effort by Stanford University and RICE University. Available at www.bakerinstitute.org/media/files/Research/36c4f094/liquefied-natural-gas-from-qatar-the-qatargas-project.pdf.

10 J. Dargin (2008), *The Dolphin Project: The Development of the Gulf Gas Initiative*, NG22. Oxford: OIES and Oxford University Press. Available at www.oxfordenergy.org/wpcms/wp-content/uploads/2010/11/NG22-TheDolphinProjectTheDevelopmentOfAGulfGasIni tiative-JustinDargin-2008.pdf.

11 Hashimoto et al., *Liquefied Natural Gas from Qatar*.

12 Ibid.

13 BP (2013), *BP Statistical Review of World Energy, June 2013*. London: BP, p. 28.

14 Tusiani and Shearer, *LNG*.

15 See https://vaaju.com/belgiumeng/qataris-would-plan-massive-invest ments-in-gas-power-plants-in-belgium.

16 See, for instance, H. Rogers (2017), *Qatar Lifts its LNG Moratorium*, OIES commentary. Available at www.oxfordenergy.org/wpcms/wp-content/uploads/2017/04/Qatar-Lifts-its-LNG-Moratorium.pdf.

17 IEA (2019), *Global Gas Security Review 2019*. Paris: IEA. Available at www.iea.org/publications/reports/globalgassecurityreview2019.

18 See www.cnbc.com/2017/06/18/qatar-wont-cut-gas-to-uae-qatar-pet roleum-ceo-says.html.

19 For a comprehensive review of recent trends and their implications for Asian importers, see IEA/KEEi (2019), *LNG Markets and Trends and Their Implications: Structures, Drivers and Developments of Major Asian Importers*. Paris: IEA/KEEi. Available at www.oecd.org/publications/lng-market-trends-and-their-implications-90c2a82d-en.htm.

20 For an analysis of these changes in LNG contracts, see T. Boersma and M. Fulwood (forthcoming). In V. Mehta (ed.), *Natural Gas in India*. New Delhi: Brookings India.

21 T. Boersma and A. Losz (2018), 'The new international political econ-omy of natural gas', in A. Goldthau, M. F. Keating and C. Kuzemko (eds.), *Handbook of the International Political Economy of Energy*

and Natural Resources. Cheltenham: Edward Elgar Publishing, pp. 138–53.

22 See, for example, www.businessinsider.com/triple-digit-oil-is-the-new-normal-2014-7?international=true&r=US&IR=T.

23 C. Ebinger, K. Massy and G. Avasarala (2012), *Liquid Markets: Assessing the Case for U.S. Exports of Liquefied Natural Gas*, Brief 12-01. Washington, DC: Brookings Institution. Available at www. brookings.edu/wp-content/uploads/2016/06/0502_lng_exports_eb inger.pdf.

24 T. Boersma, T. Mitrova, J. Typoltova, A. Galkin and F. Veselov (2018), *The Role of Natural Gas in Europe's Electricity Sector through 2030*, Working Paper. New York: Columbia Center on Global Energy Policy. Available at https://energypolicy.columbia.edu/sites/ default/files/pictures/CGEP_NaturalGasInEuropeanPowerSector_ August2018.pdf.

25 D. Sandalow, A. Losz and S. Yan (2018). *A Natural Gas Giant Awakens: China's Quest for Blue Skies Shapes Global Markets*. New York: Columbia Center on Global Energy, Policy Commentary. Available at https://energypolicy.columbia.edu/sites/default/files/file-uploads/China%20Nat%20Gas%20Commentary_CGEP_June% 202018_FINAL.pdf.

26 Y. Chen, A. Ebenstein, M. Greenstone and H. Li (2013), 'Evidence on the impact of sustained exposure to air pollution on life expectancy from China's Huai River Policy', *Proceedings of the National Academy of Sciences of the United States of America*, 110 (32): 12936–41.

27 *CBS News* (2019), 'The most polluted cities in the world, ranked': www.cbsnews.com/pictures/the-most-polluted-cities-in-the-world-ranked/51.

28 https://economictimes.indiatimes.com/news/politics-and-nation/ho pe-to-cut-air-pollution-in-100-cities-by-50-in-next-5-yrs-harsh-vard han/articleshow/62904958.cms.

29 For a recent analysis of this technology, see B. Songhurst (2017), *The Outlook for Floating Storage and Regasification Units*, NG 123. Oxford: OIES and Oxford University Press. Available at: www.oxford energy.org/wpcms/wp-content/uploads/2017/07/The-Outlook-for-Floating-Storage-and-Regasification-Units-FSRUs-NG-123.pdf.

30 T. Kott and A. Losz (2017), *They Might Be Giants – How New and Emerging LNG Importers Are Reshaping the Waterborne Gas*

Market, Working Paper. New York: Columbia Center on Global Energy Policy. Available at https://energypolicy.columbia.edu/sites/default/files/pictures/CGEPTheMightBeGiantsLNG1017_1.pdf.

31 For an analysis of the current status of floating LNG liquefaction and export terminals, see B. Songhurst (2019), *Floating LNG Update–Liquefaction and Import Terminals*, NG 149. Oxford: OIES and Oxford University Press. Available at www.oxfordenergy.org/wpcms/wp-content/uploads/2019/09/Floating-LNG-Update-Liquefaction-and-Import-Terminals-NG149.pdf?v=7516fd43adaa.

32 Some analysts have suggested that China's LNG demand, next to the greatest opportunity for LNG sellers, is also one of the greatest uncertainties in the years ahead, see: A. Losz, T. Boersma and T. Mitrova (2019), *A Changing Global Gas Order 3.0*. Columbia Center on Global Energy Policy commentary. Available at https://energypolicy.columbia.edu/sites/default/files/file-uploads/AChangingGlobalGasOrder3-0_CGEP_Commentary_041119.pdf.

33 According to the 2018 annual report of GIIGNL, in 2017 there were forty countries with installed regasification capacity. In 2018, Panama and Bangladesh joined this group, both using FSRU technology. In early 2019, Russia (Kaliningrad) joined.

34 Songhurst, *Floating LNG Update*.

35 https://energyegypt.net/end-of-an-era-the-hoegh-gallant-leaves-egypt-sets-u-s-destination.

36 J. Stern (2017), *Challenges to the Future of Gas: Unburnable or Unaffordable?* NG 125. Oxford: OIES and Oxford University Press. Available at www.oxfordenergy.org/wpcms/wp-content/uploads/2017/12/Challenges-to-the-Future-of-Gas-unburnable-or-unaffordable-NG-125.pdf.

37 For more background reading, see Tusiani and Shearer, *LNG*.

38 For details on this issue, see IGU (2011), *Guidebook to Gas Interchangeability and Gas Quality – 2011*. Available at www.igu.org/sites/default/files/node-page-field_file/Guidebook%20to%20Gas%20Interchangeability%20and%20Gas%20Quality%2C%20August%202011.pdf.

39 https://mlexmarketinsight.com/insights-center/editors-picks/antitrust/asia/jftc-report-on-lng-trade-set-to-reshape-relations-between-buyers,-sellers.

40 www.woodmac.com/news/editorial/how-four-trading-houses-are-shaking-up-the-lng-industry.

41 www.centrica.com/news/centrica-signs-mou-lng-collaboration-tok
    yo-gas.
42 For the Shell press release, see www.shell.ca/en_ca/media/news-and-
    media-releases/news-releases-2018/shell-gives-green-light-to-invest-
    in-lng-canada.html.
43 For an in-depth analysis of the challenges of market reform in East
    Asia, and what it means for natural gas pricing, see M. Fulwood
    (2018), *Asian LNG Trading Hubs – Myth or Reality*, Working Paper.
    New York: Columbia Center on Global Energy Policy. Available at
    https://energypolicy.columbia.edu/sites/default/files/pictures/Asian%
    20LNG%20Trading%20Hubs_CGEP_Report_050318.pdf.
44 A.-S. Corbeau, S. Hasan and S. Dsouza (2018), *The Challenges Facing
    India on its Road to a Gas-Based Economy*, Working Paper. Riyadh:
    King Abdullah Petroleum Studies and Research Center (Kapsarc)
    and TERI (https://doi.org/10.30573/KS--2018-DP41). Available at
    www.kapsarc.org/research/publications/the-challenges-facing-india-
    on-its-road-to-a-gas-based-economy.
45 www.crowley.com/News-and-Media/Press-Releases/Crowley-Loa
    ds-First-LNG-into-ISO-Tank-Container-at-Eagle-LNG-Partners-
    New-Plant.
46 https://uk.reuters.com/article/us-china-pollution-gas-trucks/gas-
    trucks-boom-in-china-as-government-curbs-diesel-in-war-on-smog-
    idUKKBN1CC0T0.
47 https://bioenergyinternational.com/storage-logistics/gasum-expand-
    lng-filling-station-network-sweden-norway.
48 A. Halff, L. Younes and T. Boersma (2019), 'The likely implications
    of the new IMO standards on the shipping industry', *Energy Policy*,
    126: 277–86.
49 See, for example, Stern, *Challenges to the Future of Gas*.
50 For a US case study with more nuanced considerations, see R. L.
    Kleinberg, S. Paltsev, C. K. Ebinger, D. A. Hobbs and T. Boersma
    (2018) "Tight oil market dynamics: Benchmarks, breakeven points,
    and inelasticities," *Energy Economics*, 70: 70 – 83.
51 A.-S. Corbeu and D. Ledesma (eds.) (2016), *LNG Markets in
    Transition: The Great Reconfiguration*. Oxford: OIES and Oxford
    University Press.

5 THE FUTURE ROLE OF NATURAL GAS

1 See M. Levi, M. (2015), 'Fracking and the climate debate'. *Democracy Journal*, 37: https://democracyjournal.org/magazine/37/fracking-and-the-climate-debate.

2 Carbon Tracker (2019), *Breaking the Habit: Why None of the Large Companies Are Paris-Aligned, and What They Need To Do To Get There*. London: Carbon Tracker. Available at www.carbontracker.org/reports/breaking-the-habit.

3 See www.who.int/airpollution/en.

4 World Economic Forum (2019), *The Speed of the Energy Transition: Gradual or Rapid Change?* Geneva: World Economic Forum. Available at www.weforum.org/whitepapers/the-speed-of-the-energy-transition.

5 In this context, Richard Newell in 2018 signalled that the term 'energy transition' in fact resembles more of an 'energy addition'. See www.axios.com/despite-renewables-growth-there-has-never-been-energy-transition-e11b0cf5-ce1d-493c-b1ae-e7dbce483473.html.

6 IEA (2019), *The Role of Gas in Today's Energy Transitions*. Paris: IEA. Available at www.iea.org/publications/roleofgas.

7 J. P. Banks and T. Boersma (2015), *Fostering Low Carbon Energy – Next Generation Policy to Commercialize CCS in the United State*, Issue brief 2. Washington, DC: The Brookings Institution. Available at www.brookings.edu/wp-content/uploads/2016/06/low_carbon_energy_ccs_banks_boersma_FINAL.pdf.

8 For more background, and US Senate testimony by Julio Friedmann, see https://energypolicy.columbia.edu/research/testimony/enhancing-future-ccus.

9 https://science.sciencemag.org/content/361/6398/186.

10 See R. A. Alvarez, S. W. Pacala, J. J. Winebrake, W. L. Chameides and S. P. Hamburg (2012), 'Greater focus needed on methane leakage from natural gas infrastructure', *Proceedings of the National Academy of Sciences of the United States of America*. 109 (17): 6435–40. Available at www.pnas.org/content/109/17/6435.

11 We accessed the IEA online methane tracker on 16 August 2019: www.iea.org/tcep/fuelsupply/methane.

12 Wood Mackenzie (2017), 'Upstream carbon emissions: LNG vs pipeline gas'. *Wood Mackenzie Insight*, 24 April: www.woodmac.com/news/editorial/lng-pipeline-gas-emissions.

13 IEA, *The Role of Gas in Today's Energy Transitions.*

14 For an online bibliography of research on these issues, visit: www. zotero.org/groups/248773/pse_study_citation_database/items/coll ectionKey/WEICK6IC.

15 One of the notable voluntary initiatives is the Oil and Gas Climate Initiative (OGCI), whose member companies have set a target to reduce the collective average methane intensity of their aggregate upstream oil and gas operations to below 0.25 per cent by 2025 (the baseline was set at 0.32 per cent in 2017): https://oilandgasclim ateinitiative.com/policy-and-strategy/#our-agenda.

16 For an analysis of the impact of CNG use in public transportation on local air pollutants in Delhi, see S. A. Jahilal and T. S. Reddy (2006). 'CNG: an alternative fuel for public transport', *Journal of Scientific and Industrial Research*, 65: 426–31.

17 https://www.economist.com/the-economist-explains/2018/12/13/ why-are-so-many-polish-towns-polluted.

18 J. A. De Gouw, D. D. Parrish, G. J. Fros and M. Trainer (2014), 'Reduced emissions of CO_2, NO_x and SO_2 from U.S. power plants owing to switch from coal to natural gas with combined cycle technology', *Earth's Future*, 2 (2): 75–82.

19 EIA (2018), 'Carbon dioxide emissions from the U.S. power sector have declined 28% since 2005', *Today in Energy*, 21 December: www.eia.gov/todayinenergy/detail.php?id=37816.

20 EIA (2019), 'Power sector pushed domestic U.S. natural gas consumption to new record in 2018', *Today in Energy*, 25 March: www. eia.gov/todayinenergy/detail.php?id=38812.

21 See T. Boersma and S. M. Jordaan (2017), 'Whatever happened to the Golden Age of natural gas?' *Energy Transitions*, 1 (5): https://link.spr inger.com/article/10.1007%2Fs41825-017-0005-4; J. Stern (2017), *Challenges to the Future of Gas: Unburnable or Unaffordable?* NG 125. Oxford: OIES and Oxford University Press, available at www. oxfordenergy.org/wpcms/wp-content/uploads/2017/12/Challenges-to-the-Future-of-Gas-unburnable-or-unaffordable-NG-125.pdf; and E. Stephenson, A. Douka and K. Shaw (2012), 'Greenwashing gas: might a "transition fuel" label legitimize carbon-intensive natural gas development?' *Energy Policy*, 46: 452–9.

22 The United Nations Development Programme (UNDP) annual Gap Analysis Report provides a detailed analysis of the distance between current emissions trajectories and that required to achieve various

climate goals. Available at www.unenvironment.org/resources/emiss ions-gap-report-2018.

23 The Interstate Natural Gas Association of America (INGAA) Foundation (2019), *The Role of Natural Gas in the Transition to a Low-Carbon Economy*. Report prepared by Black & Veatch Management Consulting LLC. Available at www.ourenergypolicy. org/resources/the-role-of-natural-gas-in-the-transition-to-a-lower-carbon-economy.

24 C. McGlade, M. Bradshaw, G. Anandarajah, J. Watson and P. Ekins (2014), *A Bridge to a Low-Carbon Future? Modelling the Long-Term Global Potential of Natural Gas*. London: UKERC. Available at www.ukerc.ac.uk/publications/gas-as-a-bridge.html.

25 According to the US Chemistry Council, new natural gas and NGL production has incentivized investment in new chemical manufacturing worth over US$200 billion: www.americanchemistry.com/ Shale_Gas_Fact_Sheet.aspx.

26 For transcript, see https://energypolicy.columbia.edu/research/presen tation/remarks-mayor-bloomberg-launch-columbia-universitys-cent er-global-energy-policy.

27 www.greentechmedia.com/articles/read/bloomberg-commits-500 m-to-close-all-us-coal-plants-by-2030-halt-natural-gas#gs.n5wqa0.

28 While lawmakers have agreed that more investments in CCUS projects are necessary, financial support for the legislative changes was pending, as of May 2019: www.capito.senate.gov/news/press-releases/senators-continue-push-to-issue-carbon-capture-tax-credit-guidance.

29 Oil Change International (2019), *Burning the Gas 'Bridge Fuel Myth': Why Gas Is Not Clean, Cheap or Necessary*. Washington, DC: Oil Change international. Available at http://priceofoil.org/ 2017/11/09/burning-the-gas-bridge-fuel-myth.

30 The Union of Concerned Scientists estimates that 35 per cent of existing nuclear capacity in the US is at risk of closure because it cannot compete with natural gas. This in turn would increase US power-sector emissions by 4–6 per cent. See www.forbes.com/sites/ kensilverstein/2019/05/10/if-nuclear-energy-is-replaced-by-natural-gas-say-goodbye-to-climate-goals/#b6a8b4020169.

31 For news coverage, see https://oilprice.com/Latest-Energy-News/ World-News/Argentina-Exports-First-Ever-Cargo-Of-LNG.html.

32 See the various papers on the 'decarbonisation of natural gas' by

Jonathan Stern at the OIES in the UK: www.oxfordenergy.org/authors/jonathan-stern/?v=79cba1185463.

33 W. J. E. Van der Graaff, L. Van Geuns and T. Boersma (2018), *The Termination of Groningen Gas Production – Background and Next Steps*. New York: Columbia University, Center on Global Energy, Policy commentary. Available at https://energypolicy.columbia.edu/sites/default/files/pictures/CGEP_Groningen-Commentary_072518_0.pdf.

34 For more background on Poland's relationship with the concept of 'energy security', see C. Johnson and T. Boersma (2013), 'Energy (in)security in Poland? The case of shale gas', *Energy Policy*, 53: 389–99.

35 For more on coal use and planned phase-out, see T. Boersma and S. VanDeveer (2016), 'Coal after the Paris Agreement – the challenges of dirty fuel', *Foreign Affairs*, 6 June: www.foreignaffairs.com/articles/2016-06-06/coal-after-paris-agreement; and S. VanDeveer and T. Boersma (forthcoming). In R. Falkner and B. Buzan (eds.), *World on Fire: Coal Politics & Responsibility among Great Powers*.

36 See, for example, D. M. Newbery, D. M. Reiner and R. A. Ritz (2018), *When Is a Carbon Price Floor Desirable?* EPRG working paper 1816. Cambridge University, Energy Policy Research Group. Available at www.eprg.group.cam.ac.uk/wp-content/uploads/2018/06/1816-Text.pdf.

37 www.h2-international.com/2018/09/03/h21-leeds-tests-switch-to-hydrogen.

38 J. P. Banks, T. Boersma and W. Goldthorpe (2015), *Challenges Related to Carbon Transportation and Storage – Showstoppers for CCS?* Melbourne: Global CCS Institute. Available at https://hub.globalccsinstitute.com/sites/default/files/publications/201363/Challenges%20related%20to%20carbon%20transportation%20and%20storage.pdf.

39 M. Götz, J. Lefebvre, F. Mörs et al. (2016), 'Renewable power-to-gas: a technological and economic review', *Renewable Energy*, 85: 1371–90.

40 H. Blanco, W. Nijs, J.Ruf and A. Faaij (2018), 'Potential of Power-to-Methane in EU energy transition to a low carbon system using cost optimization', *Applied Energy*, 232: 323–40.

41 C. J. Quarton and S. Samsatli (2018), 'Power-to-gas for injection into the gas grid: what can we learn from real-life projects, economic

assessments and systems modeling?' *Renewable and Sustainable Energy Reviews*, 98: 302–16.

42 C. Van Leeuwen and M. Mulder (2018), 'Power-to-gas in electricity markets dominated by renewables', *Applied Energy*, 232: 258–72.

43 H. Blanco and A. Faaij (2018), 'A review of the role of storage in energy systems with a focus to Power to Gas and long-term storage', *Renewable and Sustainable Energy Reviews*, 81: 1049–86.

44 T. Van Melle, D. Peters, J. Cherkasky, R. Wessels, G. U. R. Mir and W. Hofsteenge (2018), *Gas for Climate – How Gas Can Help to Achieve the Paris Agreement Target in an Affordable Way*. Utrecht: Ecofys. SISNL17592. Available at www.gasforclimate2050.eu/files/files/Ecofys_Gas_for_Climate_Feb2018.pdf.

45 N. Scarlat, J.-F. Dallemand and F. Fahl (2018), 'Biogas: Developments and Perspectives in Europe', *Renewable Energy*, 129: 457–72.

46 S. Mittal, E. O. Ahlgren and P. R. Shukla (2018), 'Barriers to biogas dissemination in India: a review', *Energy Policy*, 112: 361–70.

47 For example, Heidelberg Cement is constructing the world's first zero-emissions cement plant in Norway, see www.euractiv.com/section/energy/news/worlds-first-zero-emission-cement-plant-takes-shape-in-norway.

48 C. McGlade, S. Pye, P. Ekins, M. J. Bradshaw and J. Watson (2018), 'The future role of natural gas in the UK: a bridge to nowhere?' *Energy Policy*, 113: 454–65 (https://doi.org/10.1016/j.enpol.2017.11.022).

49 Committee on Climate Change (2019), *Net Zero – The UK's Contribution to Stopping Global Warming*. London: CCC. Available at: www.theccc.org.uk/publication/net-zero-the-uks-contribution-to-stopping-global-warming.

50 J. Stern (2019), *Narratives for Natural Gas in Decarbonising European Energy Markets,* NG 141. Oxford: OIES and Oxford University Press. Available at www.oxfordenergy.org/publications/narratives-natural-gas-decarbonising-european-energy-markets/?v=79cba1185463.

51 In recent months, landmark decisions have been announced in California, but also Indiana, where state regulatory authorities argued that utilities' proposed investments in gas-fired electricity-generation capacity were not well founded, considering the alternative of investing in low-cost renewable energy, in combination with electricity storage technology.

52 www.citylab.com/solutions/2016/11/amsterdam-natural-gas-ban-20
   50-climate-change-regulations/508022.

6 The Golden Age of Gas?

1 IEA (2011), *Are We Entering a Golden Age of Gas?* Paris: IEA.
   Available at www.iea.org/publications/freepublications/publication/
   WEO2011_GoldenAgeofGasReport.pdf.
2 Details of the conference and various presentations can be found at
   www.lng-conference.org/english.
3 IEA (2019), *Gas 2019: Analysis and Forecast to 2024*. Paris: IEA.
   Available at www.iea.org/gas2019.

Postscript

1 www.iea.org/commentaries/the-coronavirus-crisis-reminds-us-that-
   electricity-is-more-indispensable-than-ever.
2 www.thenational.ae/business/economy/what-the-energy-industry-wi
   ll-look-like-after-the-coronavirus-1.992625.
3 https://uk.reuters.com/article/uk-health-coronavirus-exxon-mobil-
   mozamb/exclusive-coronavirus-gas-slump-put-brakes-on-exxons-
   giant-mozambique-lng-plan-idUKKBN21801A.
4 www.carbonbrief.org/analysis-coronavirus-has-temporarily-red
   uced-chinas-co2-emissions-by-a-quarter.
5 www.oxfordenergy.org/wpcms/wp-content/uploads/2020/03/Geo
   political-shifts-and-Chinas-energy-policy-priorities.pdf?v=7516fd43
   adaa&utm_source=newsletter&utm_medium=email&utm_campai
   gn=newsletter_axiosgenerate&stream=top.
6 www.nytimes.com/reuters/2020/03/17/world/europe/17reuters-he
   alth-coronavirus-poland-ets.html.
7 For a coherent argument against the comparison between climate
   change and COVID-19, see https://thebreakthrough.org/issues/ene
   rgy/covid-19-climate.

Selected readings

Each chapter is supported by detailed notes that provide access to the source materials used in our analysis. Here we focus on the more substantial treatments of the issues covered in each chapter, and the organizations that provide information on developments in the global gas industry.

Chapter 1 discusses the fundamentals of natural gas. How is natural gas produced, processed and transported? What do we typically use natural gas for? Where do we find the resource? How do we trade it? Does it have a place in an increasingly GHG-constrained economy? And, for this book, what political economic lens do we offer to examine further this resource, which is the world's fastest-growing fossil fuel? For a discussion of the natural gas industry, see Vivek Chandra's *Fundamentals of Natural Gas: An International Perspective*, 2nd edition (Tulsa, OK: PennWell Publishers, 2017); Vaclav Smil's *Natural Gas: Fuel for the 21st Century* (Wiley: Chichester, 2015) and Thierry Bros' *After the US Shale Gas Revolution* (Paris: Éditions Technip, 2012). For a recent book-length analysis of the geopolitics of natural gas, see Agnia Grigas' *The New Geopolitics of Natural Gas* (Cambridge, MA: Harvard University Press, 2017). Statistics on the natural gas industry can be found on the websites of the International Gas Union (IGU) (www.igu.org); the International Group of Liquefied Natural Gas Importers (GIIGNL) (https://giignl. org); and the major oil and gas companies, particularly BP's

Statistical Review of World Energy and its *Energy Outlook* (www.bp.com/en/global/corporate/energy-economics.html). The IEA also makes available reports and statistical materials; some are free, such as its annual gas security report, but most, including its annual gas market report that includes a five-year forecast, are behind a paywall. Institutions with a subscription to the OECD's i-Library should have access to the IEA's materials. We also rely heavily on the outputs of two think tanks that produce research reports on the natural gas industry: the Gas Programme at the OIES (www.oxfordenergy.org), and the Columbia University Center on Global Energy Policy (https:// energypolicy.columbia.edu/topics/natural-gas). Other think tanks that publish on the geopolitics of natural gas include: the Center for Strategic and International Studies (www.csis. org) and the Brookings Institution (www.brookings.edu), both in Washington, DC, and the Clingendael International Energy Programme (www.clingendaelenergy.com) in the Netherlands.

Chapter 2 investigates international natural gas trade by pipeline, and how that by default created interdependencies between sellers, buyers and sometimes middlemen – at times with serious geopolitical consequences. The chapter identifies that the overwhelming majority of natural gas that is produced does not cross an international border, and that, increasingly, the traditional way of shipping natural gas is under pressure, with the rise of LNG, and demand growth in emerging economies that often lack elaborate pipeline systems. For a discussion of the history of gas pipelines between Russia and Europe, the key source is Per Högselius' *Red Gas: Russia and the Origins of European Energy Independence* (Basingstoke: Palgrave Macmillan, 2012). Jeff Makholm's *The Political Economy of Pipelines* (University of Chicago Press, 2012) provides a detailed historical analysis of the politics and economics of oil and gas pipelines in the US. James Henderson and Simon Pirani's edited volume draws on the expertise of the OIES to provide an excellent analysis of the interplay between economics and geopolitics in *The Russian Gas Matrix: How Markets Are Driving Change* (Oxford: OIES and Oxford University Press, 2014); while Andreas Goldthau and Nick Sitter examine the political economy of the EU's energy policy in *A Liberal Actor in a Realist World: The European Union Regulatory State and the Global Political Economy of Energy* (Oxford University Press, 2015). Tim Boersma's *Energy Security and Natural Gas Markets in Europe: Lessons from the EU and the United States* (London: Routledge, 2015) offers a somewhat different perspective by looking at EU gas security through the lens of US experience. Detailed studies of natural gas in Central Asia are sparse and a key source is the collection edited by Andreas Heinrich and Heiko Pleines: *Export Pipelines from the CIS Region: Geopolitics, Securitization and Political Decision-Making* (Stuttgart: Ibidem-Verlag, 2014). The Russia–China energy relation is attracting greater attention, and Bobo Lo's book *Axis of Convenience: Moscow, Beijing and the New Geopolitics* (London: Royal Institute of International Affairs, 2008) provides the historical context, while the details of the energy relationship are documented in Keun-Wook Paik's *Sino-Russian Oil and Gas Cooperation: The Reality and Implications* (Oxford: OIES and Oxford University Press , 2012). Access to more recent analyses is provided via the notes for chapter 2.

Chapter 3 discusses the rise of what is often referred to as 'unconventional gas production', including natural gas trapped in shale-rock layers. It discusses the unique circumstances that allowed for the shale revolution to take place in the United States, and the dramatic consequences it has had on global energy markets and geopolitics. It then continues with various case studies examining the possibility of 'the revolution' spreading to other parts of the world. The shale revolution in the US has spawned a large number of books examining the phemomena from a variety of perspectives. We have shied away from the more popularist and polemical and suggest: Trevor Houser and Shashank Mohan's *Fueling Up: The Economic Implications of America's Oil and Gas Boom* (Washington, DC: The Peterson Institute, 2014); Rusty Braziel's *The Domino Effect: How the Shale Revolution Is Transforming Energy Markets, Industries and Economies* (Houston: NTA Press, 2016); Daniel Raimi's *The Fracking Debate: The Risks, Benefits, and Uncertainities of the Shale Revolution* (New York: Columbia University Press, 2018) and Meghan O'Sullivan's *Windfall: How the New Energy Abundance Upends Global Politics and Strengthens America's Power* (New York: Simon and Schuster, 2017). We have also drawn from a number of academic edited collections and recommend: R. Quentin Grafton, Ian G. Cronshaw and Michal C. Moore's *Risks, Rewards and Regulation of Unconventional Gas: A Global Pespective* (Cambridge University Press, 2017) and Shanti Gamper-Rabindran's *The Shale Dilemma: A Global Perspective on Fracking & Shale Development* (University of Pittsburgh Press, 2018).

Chapter 4 addresses the rise of LNG as an increasingly popular way to transport the resource. It provides a brief history of the market for LNG, and discusses how this once niche market has been growing rapidly, allowing an ever increasing number of countries to start importing natural gas. The chapter also discusses that the maturing of a market can be spurred by single events (e.g. the nuclear disaster in Fukushima), but that coming of age does tend to take time, and further market reform in various key markets in Asia could likely support LNG becoming a more global commodity (like oil or coal) at some point in the future. There are two annual industry reports that provide essential statistics on the global LNG industry. The first is the Annual Report of the International Group of LNG Importers (https://giignl.org) and the second is the Annual LNG Report of the IGU (www.igu.org). Shell (www.shell.com) produces an annual LNG Outlook that presents the views of one of the key actors in the industry. There are also numerous industry events, publications and reports on the LNG industry, but most of these are only available on a commercial basis. One standard reference on LNG that we would recommend is Michael Tusiani and Gordon Shearer's *LNG – Fuel For a Changing World – A Nontechnical Guide, 2nd edition* (Tulsa, OK: PennWell Publishers, 2016); in addition, a recent and very comprehensive analysis of current developments in the industry is found in the OIES volume edited by Anne-Sophie Corbeau and David Ledesma: *LNG Markets in Transition: The Great Reconfiguration* (Oxford: OIES and Oxford University Press, 2016).

Chapter 5 turns to the role of natural gas as countries around the world get set to transition to lower-carbon energy systems. It observes that there are major discrepancies between various geographies around the world when it comes to policy narratives about the role that natural gas may play. Broadly, in the OECD world, increasingly, questions are asked about the compatibility of natural gas with scenarios of deep decarbonization, even though in most cases, and certainly in the United States, natural gas demand has grown substantially in recent years. In non-OECD developing countries, concerns about providing energy access and improving local air quality typically trump concerns about reducing GHG emissions, and natural gas demand consequently is anticipated to grow significantly in the years ahead (and, in the process, GHG emissions can often be reduced, as natural gas replaces more-polluting fuels such as coal and oil). The chapter outlines the environmental debates, and uncertainties, which are destined to grow and become key policy concerns in the years ahead. This is a fast-moving topic and most of the research and discussion can be found in the academic journal literature, and reports of international organizations, environmental NGOs and think tanks, much of which is detailed in the notes for chapter 5. Christophe McGlade and colleagues at the UKERC (www.UKERC.ac.uk) provide an early analysis of the potential future role of natural gas in *A Bridge to a Low-Carbon Future? Modelling the Long-Term Global Potential of Natural Gas* (London: UKERC, 2014). More recently, a series of reports by Jonathan Stern at the OIES, including *Challenges to the Future of Gas: Unburnable or Unaffordable?* NG 125 (Oxford: OIES, 2017), provides the best entry point into the debate about the future role of natural gas and the challenge of decarbonization. The IEA has also joined the debate and published a free-to-access report on *The Role of Gas in Today's Energy Transitions* (Paris: IEA, 2019). Opposing views on the future for gas are found in various industry outlooks, such as BP's *Energy Outlook* and Equinor's *Energy Perspectives*, on the one hand, and the reports of environmental NGOs such as Oil Change International's *Burning the Gas 'Bridge Fuel Myth': Why Gas Is Not Clean, Cheap or Necessary* (Washington, DC: Oil Change International, 2019) and Carbon Tracker's *Breaking the Habit: Why None of the Large Companies Are Paris-Aligned, and What They Need To Do To Get There* (London: Carbon Tracker, 2019), on the other hand.

Index